AMBER

The Golden Gem Of The Ages

AMBER

The Golden Gem Of The Ages

Patty C. Rice, Ph.D.

VNR **VAN NOSTRAND REINHOLD COMPANY**
NEW YORK   CINCINNATI   ATLANTA   DALLAS   SAN FRANCISCO
LONDON   TORONTO   MELBOURNE

Van Nostrand Reinhold Company Regional Offices:
New York Cincinnati Atlanta Dallas San Francisco

Van Nostrand Reinhold Company International Offices:
London Toronto Melbourne

Manufactured in the United States of America

Published by Van Nostrand Reinhold Company
135 West 50th Street, New York, N.Y. 10020

Published simultaneously in Canada by Van Nostrand Reinhold Ltd.

15 14 13 12 11 10 9 8 7 6 5 4 3 2 1

Library of Congress Cataloging in Publication Data

Rice, Patty C
  Amber, the golden gem of the ages.

  Includes index.
  1. Amber. I. Title.
QE391.A5R52   553'.29   79-22189
ISBN 0-442-26138-1

Preface

As a collector of amber, I found a broad gap in the current literature about amber, as well as a lack of information on amber among dealers of gems and minerals and of antiques. Several unfortunate experiences in buying "amber" beads from reputable dealers, only to later discover the material was an imitation, encouraged me to research the subject in depth.

This was no easy task. At the beginning I found only short articles on amber in various encyclopedias and other reference books. Much to my dismay, most books published in English on amber are now out of print and extremely difficult to locate, and when I found one, it was disappointing to learn the information was outdated.

Detecting imitation amber can be difficult when so many substitutes are on the market. They are sold in good faith by dealers as "antique amber," "cherry amber" or "African amber," and command prices comparable to those of true amber. For example, claimed to be "antique cherry red amber," bakelite plastic necklaces recently sold for prices as high as $160 to $200 in antique shops, when their true value lay within the $35 to $50 range. Yellow, orange and brown "African amber," or "copal" beads, are sold at $5 to $7 each in bead shops around the country. When tested, the majority of these proved to be modern plastics.

This book is based on my own collecting endeavors as well as on my jewelry work using amber, and is an attempt to enlighten both the collector and lapidary in all aspects of the science of amber, including its history, its lore, its nature, its modern sources and its commerce.

PATTY C. RICE, PH.D.

Acknowledgments

Acknowledging all who helped make this book possible is a difficult task. Many have guided, inspired, offered suggestions, shared their wisdom and collections of amber with me and criticized the manuscript.

I particularly acknowledge and thank my dear husband, who provided much assistance in all stages of the work. It was through his encouragement and support that I developed my own amber collection and gained the courage to produce this book. His expertise in photography, drawing and proofreading was of great assistance.

I am especially grateful to John Sinkankas for opening his home and making his rare book collection available to me for extensive research which would not have been possible otherwise. His valuable suggestions on the technical aspects relating to amber as a gem were extremely helpful, and I am indebted to him for his guidance.

Since most published works on amber are found only in rare book collections, I relied much on the library services of Cora Ellen DeVinney and Dick Palmer for obtaining books from libraries throughout the United States. Their services are most appreciated.

Many of the works on amber are written in German, Lithuanian and Russian. The following friends gave much assistance in translations of important information from foreign publications: Stephanie Kaunelis, Hans Breiling, David Odenwalder, Zophia Ladak, J. Brodzinsky and John Sinkankas. I thank them for their contributions.

Amber Forever, Amberica, Agneta Baltic Amber, Giovanni Creations, Gifts International, Gifts of Scandia and Kuhn's Baltic Amber, all merchants dealing in amber, graciously answered questions and allowed their amber collections to be photographed. I thank them, as well as the many friends who permitted me to bring my photographic equipment into their homes—which often involved carrying several pounds of equipment, such as floodlights, tripod, copystands and background drops—so their amber collections could be photographed. Their tolerance during this process was remarkable and greatly appreciated.

I also thank Dr. Charles Lewis, Annabel McIlnay and Vic Dene, who provided useful critiques of the manuscript.

I express my appreciation to Bob Brison for artwork; to Pat Hall for technical assistance with photography; and to Peggy Koscielniak, who spent many hours typing the manuscript.

Finally, I want to thank the many mineral and gem club members who viewed my collection and encouraged me to complete the book so answers to their questions about amber might be recorded.

PATTY C. RICE, PH.D.
MT. CLEMENS, MICHIGAN

Contents

AMBER

The Golden Gem Of The Ages

Part 1

Amber,
The Golden Gem
of the Ages

1
An Introduction

Eurymachus
Received a golden necklace, richly wrought,
And set with amber beads, that glowed as if
With sunshine.

Homer

Amber, a gemstone sought after by ancient Stone Age sun worshippers because its beautiful radiance resembled the sun's rays, well deserves the title, "golden gem of the ages." In the civilizations of the early Greeks and Romans, amber was so revered it was available only to nobility. Ladies of the Roman court desired it for its brilliant hue and for the protection from evil spells which it was believed to bestow upon the wearer. In reverence to its talismatic powers, gladiators wore amber amulets when venturing into the Coliseum. Throughout Europe, amber was worn as protection against various and sundry illnesses.

Although ancient man and the peoples of many later civilizations treasured amber as highly as gold, little was known of its origin until the age of science brought proof that it originated from the sticky resin which flowed from prehistoric trees. Few gems match amber in respect to its mode of creation, the depth of its history and its transmission of aesthetic pleasure to man. None can match it in the range of human knowledge and scientific information its study reveals.

Over the centuries, amber primarily came into the hands of man from the seashores and outcroppings of amber-bearing strata near the Baltic Sea; however, small deposits have been found in other places throughout the world. By the late 1800's, the Baltic amber industry had become highly organized, with extensive mining taking place in East Prussia, a region which is now part of Poland, the Kaliningrad Oblast of the USSR and the Lithuanian SSR. Over a million kilograms of amber were produced in this region alone between 1895 and 1900, and much is still being produced in the Baltic (and, to a lesser extent, elsewhere).

During the 1800's, scientists began studying the insects and other evidences of fauna and flora of the past entrapped in amber resin as it flowed from trees in primeval forests in the "Amberland" of the Baltic. An excellent summary (which is now difficult to find) of all aspects of amber is the work published in 1932 by George C. Williamson.[1] Since Williamson's study, research on amber has tended to focus on the study of fossil inclusions, the geology of the amber-bearing beds and (especially) the genetic relationships between the ancient fossilized resins and those being produced today by living trees. Sophisticated techniques, such as infrared spectroscopy, X-ray diffraction, mass spectroscopy and gas-liquid chromatography, are being used today. They provide paleobotanists with a better understanding of the evolution of some present day vegetation.[2]

Thus, amber is valued more than ever—not only by connoisseurs and collectors, but also by paleobotanists and paleozoologists, not to mention geologists, archaeologists and anthropologists, who find in amber an excellent means of tracing the development of our lands and their inhabitants in the distant past.

AMBER AS A GEMSTONE

Rough or "block" amber is commonly found in irregular lumps or rounded nodules, also in grains, drops and stalactitic masses, reflecting the shapes initially assumed by outpourings of tree resins. Pieces are generally small, weighing up to about 200 grams, although considerably larger masses—in excess of several kilograms—have been found. The largest pieces on record weighed from 10 to as much as 20 kilograms.[3] Of all the amber produced, however, only 5 to 6 percent of the pieces measure 30 millimeters, or a little over one inch, in diameter. But regardless of size, virtually all amber is now put to good use.

Amber is often collected directly from the waters of the Baltic Sea since it is just buoyant enough to float in salt water. Typically, such floating amber receives a semi-polish because of the action of waves and beach sands, which pummel these so-called "sea-stones" until they are divested of their natural crust and smoothed on all surfaces. (See Color Plate 1 for an example of sea-stone.) In contrast, mined amber, or "pit-amber," is covered with a dark brown crust that must be removed before one can discover the quality of the material within. (See Color Plate 2 for an example of pit-amber.) Major production of amber in the Baltic region is *not* from the sea, but from mining in the earth, with the largest open-cut amber mining operation in the world located in the Kaliningrad Oblast SSR.[4]

Because amber is fossilized tree resin, it is resinous in luster and lacks any crystalline structure such as is so commonly found in mineral gemstones. It otherwise resembles what one would expect from a hardened tree resin. Some recent X-ray work on amber does show that crystalline organic compounds are present. These structures are thought to be crystals of succinic acid or possibly other compounds.[5]

Fig. 1-1. Rough Baltic amber pieces.

Fig. 1-2. Large amber lump weighing about 1 kilogram.

Amber is commonly yellow to honey-colored, but within the range of these hues are many subtle variations, as depicted in Color Plate 3. In fact, the hues range from light yellow to dark brown and include lemon yellow, orange, reddish brown and hues that are almost white. Some are so pale the amber seems colorless. Even greenish and bluish tints, and, more rarely, violet tinges are found. Approximately 250 different color varieties are known, although by far the most common hues are yellow, orange and brown. White pieces with a slight yellowish tinge or colored like bone are rare, but perhaps the rarest of all are the reddish, bluish and greenish hues.

In addition to variations of color, amber can be absolutely transparent or completely cloudy, and variations can often be found in all ranges within one specimen. The various colors are commonly caused by light interference on minute gas bubbles, or, less often, by stains of minerals or compounds resulting from the decomposition of organic debris enclosed in the amber. It is believed that reddish hues are the result of oxidation. If amber is exposed to a radioactive substance for a month or more, its color darkens.

The red color caused by oxidation seldom penetrates completely. For example, the amber found in Stone Age tombs, usually encrusted with a reddish brown layer, is shown to be still yellow inside when it is cut through. In

Fig. 1-3. Large amber lump with crust showing. *(Courtesy Gifts International, Chicago.)*

modern times, this "aging" can be produced artificially by heating amber at low temperature in an oven. By this means, a rich antique reddish brown hue is induced.

Amber is brittle and breaks with a conchoidal fracture surface which has a curved appearance, often with numerous concentric circular ridges. Large, clear pieces of amber often contain characteristic irregular fractures or cracks within them and suchs flaws or "feathers" are not considered to detract from the piece's value. Some clear masses were formed from concentric layers of resin, one with another, like the layers of a hailstone, and are called "shelly" amber.[6] Fractures in this type of amber are likely to follow the concentric layering. Other lumps are compacted in a single mass and, therefore, are termed "massive" amber. Specimens have been found representing all variations and gradations, from shelly to massive, and such structures are important in determining the manner in which any piece of rough amber can best be utilized.

While amber is soft enough to be cut with a pocket knife, it is scarcely scratched by a fingernail. When scraped with a knife, it tends to crumble into a powder. Parings or thin peels cannot be obtained.

When chips of amber are burned, a smoke with a pleasant resinous odor of pine is emitted, and for this reason amber was burned as incense in temples in the Far East. If a lump of amber is rubbed vigorously, not only does it give off a piney scent, but the friction causes it to produce enough negative static electrical charge to pick up small particles of tissue paper. The word *electricity* was actually derived from *elektron,* the Greek name for amber. Unlike other mineral gemstones, amber is warm to the touch, since it is a poor conductor of heat.

Fig. 1-4. Large lump of shelly amber containing a swarm of insects; weight over 1 kilograms. *(Courtesy Kuhn's Baltic Amber, Florida.)*

In the past, Baltic amber was considered by most gemologists and mineralogists to be the only "true amber." Baltic amber is the most abundant of all ambers and is more often used for ornamental purposes since it is generally harder than other fossil resins. However, amber from the Dominican Republic is becoming available commercially and is crafted into beautiful ornaments by the local craftsmen.

Another name given by mineralogists to Baltic amber is *succinite,* a term based on the chemical analysis showing Baltic amber to contain at least 3 to 8 percent succinic acid, more than the amount found in any other comparable fossil resin. For this reason, the presence of succinic acid provides an important clue in identifying Baltic amber, and its presence or absence is used to specify the probably places of origin of amber articles found in Stone Age tombs.[7]

Modern studies of the botanical origin of resins also use succinic acid content for identifying varieties of amber. Based on its presence, amber resins have been placed in two classes: the *succinites,* those containing succinic acid, and the *retinites,* those which do not contain succinic acid. Recent studies have shown that the natural formation of amber—that is, its production by resinous prehistoric trees—was similar in all regions, and that while the *formation* may have taken place in different epochs, the *processes* were comparable. Therefore, the presence or absence of succinic acid represents only a variation in the chemical composition and not the existence of a distinctly different formation process.

Up to 1930, the largest production of amber was from deposits in the Baltic area, which today comprises the coastal regions of Denmark, Sweden, Northern Germany, Poland and the Soviet Union. The Danish amber belt lies on the western shore of the Baltic Sea and includes the Frisian Islands along the coastline of the North Sea. In Poland, amber occurs along the entire Baltic coast, but most abundantly in the region of Gdańsk. Amber from the USSR is mainly produced along the Baltic coasts of Lithuania, with small amounts found in Latvia and Estonia. The Kaliningrad Oblast, where the largest concentration of amber occurs, produces approximately two-thirds of the world's supply today (see Fig. 1-5). This area is unique not only for its quantity of amber, but for the variety of types of amber found.

The countries which import the most amber from the Soviet Union are West Germany and Japan. Most Baltic amber sold in the United States is imported either from West Germany or Scandinavia, inasmuch as these countries have established commercial sources of supply, and are easily accessible to individual United States dealers. After World War II, the Kaliningrad area mine was reported to have been closed by the Russians (however, it is now re-opened). The commercial exporter for amber from this region is the Almazjuvelirexport of Moscow, whose catalog contains natural amber products of all shades within the yellow range and of a variety of textures, as well as "fired" natural amber and pressed amber. The firing and pressing processes for utilizing small chips and for enhancing the beauty of natural amber will be described in detail in later chapters.

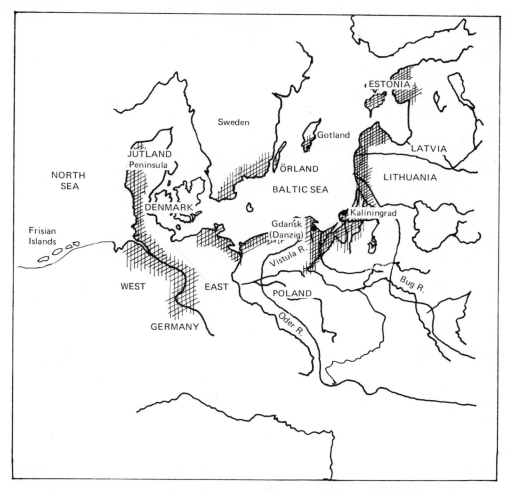

Fig. 1–5. Occurrence of Baltic Amber in various locations along the Baltic Coast.

Elsewhere, small quantities of amber or fossil resins have been found in Sicily, Romania, China, Burma, Thailand (Siam), Japan, the Soviet Union, Canada, Mexico and the United States. Currently, large quantities are being mined in the Dominican Republic. These sources produce fossil resins which differ only slightly in chemical composition from Baltic amber.

More than 20 fossil resins, similar to Baltic succinite, but varying in geological age, some physical and chemical properties and perhaps botanical source, have been found in numerous localities throughout the world. Most have little commercial value; but those used as gem material will be included in the chapter on varieties of fossil resins.

As evidence of an era of craftsmanship from the past, we frequently find carefully faceted beads of clear, transparent golden-colored amber in estate sales or antique shops. These older styles of amber beads are often a golden orange color because of exposure over the years to the atmosphere. As has

Fig. 1-6. Microphotograph of crazed effect on the surface of antique faceted bead.

been pointed out, amber, after being polished, can deepen in color as it is exposed to the atmosphere over a period of time. Beads fashioned during the early part of this century or before are often slightly crazed on the surface. This is thought to result from long-sustained variations in temperature. Amber also tends to craze when it is subjected to extreme heat because of its low melting point and the escape of volatile substances. Crazing is an important feature to look for when examining old beads to determine whether they are true amber or imitation. (See Fig. 1-6.)

Another feature useful in distinguishing genuine faceted amber beads from imposters is that facets in the gem material tend to show signs of wear because of the gem's relative softness. The formerly sharp junctions between the facets may be rounded. The holes through the beads also may be slightly chipped or irregular at their ends, resulting from constant rubbing of the stringing material and self-abrasion of the beads. Interestingly, the aging process adds to the mellow beauty and the warm, soft glow of the stones. (See Fig. 1-7.)

The antique trade has coined a variety of picturesque names for the many variations in amber beads based on color and transparency. Names such as "cherry," "tomato," "egg yolk," "goose fat," "root" and "antique" are often used to describe these differing characteristics. The peasants of Poland have names for over 200 variations of amber.

The Baltic amber industry itself classifies types of amber and provides specific names for the jewelry trade. These classifications are based on variations in color shades as associated with different degrees of transparency. There are two main categories: *clear* or *transparent* amber and *cloudy* amber. Cloudy amber is found in all degrees, from semi-transparent to opaque. The trade further separates cloudy amber into "fatty" or

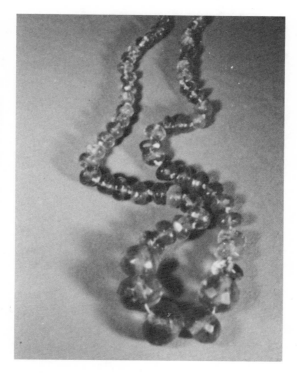

Fig. 1-7. Old faceted beads with prehistoric insects enshrined within several of the beads.

"flohmig," "bastard," "semi-bastard," "bone" or "osseous" and "foamy" or "frothy" amber. In actuality, many combinations of differing diaphaneity are often found within the same piece of amber. All these varieties are used in making art objects and jewelry, with each type appealing to connoisseurs in various areas of the world at different times.[8] Furthermore, each type has different capabilities for acquiring a polish, and it is this quality that dictates the use to which the pieces will be put in jewelry or artistic works.

Clear Amber

Clear or transparent amber takes a high polish and is in great demand for making faceted beads. Since the era of the amber guilds of the Middle Ages, amber beads—as meticulously faceted as any other transparent semi-precious gem—were produced in great quantities by skilled craftsmen. During the 1920's, when amber was second only to diamonds in the United States gem imports,[9] much of the clear amber imported was in the form of faceted beads, and these can still be found in antique shops. The United States importers tended to favor clear, golden varieties, and this is the hue most often thought of when speaking of amber. (See Color Plates 4, 5 and 6.)

Fig. 1-8. "Loleta" style necklace of clear tumbled amber nodules.

The clear or partially clear shelly amber, with its loosely adhered layers, is rarely cloudy throughout and seldom contains cloudy layers as does massive amber. Both the structure and the resulting diaphaneity of the amber are caused by the formation process at the time the amber-producing resin was exuded from the prehistoric trees.[10] Transparent specimens range in hue from almost colorless to dark reddish yellow. The "water-clear" amber is very rare. Most frequent are "yellow-clear" colors.

Cloudy Amber

Cloudy ambers are highly prized in Europe for their muted beauty. The natural shape of the rough gem and the extent of its cloudiness are considered when pendants and beads are made. Fatty or flohmig amber is slightly turbid and appears to have a fine dust suspended within the pieces (see Color Plate 7). It takes a high polish and provides beautiful, glistening gems with the appearance of whipped honey. Fatty amber is semi-transparent but not as turbid as the cloudy bastard amber described below. When viewed under a microscope, a thin slice of fatty amber reveals an abundance of tiny bubbles, which dilute the basic color and produce the clouded appearance. The East Prussian word, "flohmfet," refers to the translucent yellowish fat of

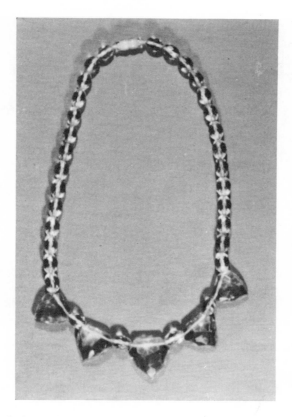

Fig. 1–9. Exquisitely faceted clear amber beads from the late 1800's. *(Courtesy Kuhn's Baltic Amber, Florida.)*

the goose. Thus, the term fatty amber is descriptive of the actual appearance of the amber.[11]

Cloudy bastard amber takes a good polish, but is more turbid than fatty amber (see Color Plate 8). Again, the cloudy appearance is caused by a multitude of small bubbles within the mass. This variety of amber is common, and there are several degrees of cloudiness. Lumps that are almost clear can sometimes be clarified by heating the piece gently in an oil such as rapeseed.* Some pieces may have portions which display swirling clouds in a clear ground mass (see Color Plate 9). In addition to the various degrees of cloudiness, variations in color also appear. Pieces may be whitish, yellowish, brownish yellow or even reddish brown. The white or grayish white pieces are referred to as pearl-colored. Paler tones are misleadingly called "blue amber" (although there *is* a rare amber of actual blue color, as can be seen in Color Plates 10 and 11).

The so-called "egg yolk" amber is a yellow or brownish yellow bastard amber within a clear ground mass. In Europe it is called kumst-colored. The work "kumst" in the East Prussian dialect of the German language means cabbage or sauerkraut.[12] (See Color Plates 12 and 13.)

* See the chapter on preparation and working of amber.

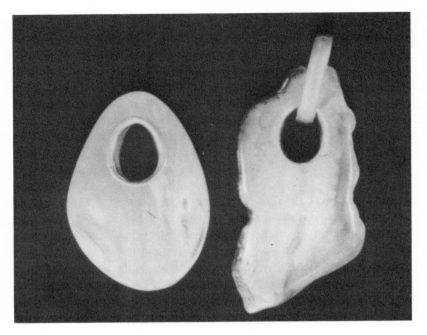

Fig. 1-10. Fatty amber pendants from Poland. *(Courtesy Amber Forever, Florida.)*

Semi-bastard amber is between bastard and osseous amber (described below) in terms of diaphaneity. It appears similar to the osseous amber but has the capability of receiving a polish similar to that of the bastard amber. The antique trade sometimes labels this kind of amber, which has deepened in color from the process of aging, as "tomato" amber (see Color Plate 14).

Bone or osseous amber is an opaque variety which is slightly softer than the others and consequently does not polish as well. The color is whitish yellow to brown and the specimens give the general appearance of bone or ivory. The pieces can be variegated with "osseous-clear" or "osseous-bastard." (See Color Plate 15 for an example of bone amber.)

The last variety in order of usefulness to the jewelry industry is the foamy or frothy amber, being opaque, very soft and incapable of receiving a polish. Occasionally, large, interesting lumps are used to form primitive and unusual pendants. Pieces are also used in amber mosaics.

The rarest color varieties of amber are green and blue, followed by pearl-colored amber and that called kumst. Because of their scarcity in Europe, these are generally favored over clear amber and fetch a higher price in most European markets. In contrast, clear amber and its varieties with petal-like inclusions are currently most popular in the United States.

In the past, it was thought that differences in turbidity in amber were caused by the presence of varying amounts of minute droplets of water. However, modern microscopic examination shows turbidity results from a vast number of microscopic gas bubbles enclosed within the gemstone. Between them, the resin is clear.

Bubble inclusions interfere with light passage through the amber mass, resulting in dilutions of color, variations in color and turbidity. Even the hue in the rare blue, green and grayish green varieties of amber results from such enclosed bubbles. (See Color Plate 16 for an example of green amber.) Green amber is thought to have formed in marshy areas where decaying organic material may have influenced the color. In the case of blue amber, the hues range from azure to sky-blue. When blue amber is heated over a period of time in an oil with the same refractive index as amber, the piece clarifies but the color changes to an ordinary yellow hue.[13] Both blue and green ambers are generally turbid.

Inclusions in amber, both inorganic and organic, also influence the color. In foamy amber, pyrite is commonly enclosed in the form of thin lamellae which fill cracks, particularly in the layered shelly amber. If too much pyrite is present, it interferes with the lapidary treatment of the piece.[14]

Black-and-white amber, or a kind of osseous amber dotted with black inclusions, is valued for its rarity. The black inclusions are organic material, generally decayed botanical debris. (See Color Plate 17 for an example of black-and-white amber.) Finely divided particles of carbonized wood also appear black in color and sometimes are present abundantly in amber formed in boggy or marshy areas. Amber of this kind retains its typical coloring, within the yellow range, but the carbonized plant debris often appears as black specks suspended in the amber. Paleobotanists believe that particles of decayed wood found enclosed in amber are probably the remains of the same coniferous trees which exuded the original resin millions of years ago.[15]

During the 1920's, amber bead necklaces with an insect embedded in each bead were especially in vogue. A fine example of a necklace from this period may be seen in the Hall of Gems of the Field Museum of Natural History in Chicago. Insect inclusions in Baltic amber are rare, but those that exist show the insects to be perfectly preserved and complete in every detail, including wings, antennae and legs. (There will be further discussion of the flora and fauna found in amber in the chapter on scientific studies of amber.)

AMBER AS A FOSSIL RESIN

Amber is one of the few gemstones of organic, rather than mineral, origin. Essentially, amber is a fossilized resin from prehistoric evergreens or other now-extinct species of resin-producing trees which flourished in large forests as far back as 50 million years ago. Based on the geological calendar, from the Eocene epoch to as late as the Miocene epoch of the Tertiary period, the northern parts of Europe (including regions now encompassing Finland, Scandinavia, the Baltic Sea and northern Russia) are thought to have been one continuous land mass, labeled by geologists as the pre-Fennoscandian continent (see Table 1-1). It is believed to have extended to as far as Spitsbergen to the north and Iceland to the west, and to have covered parts of Greenland and North America, the British Isles and northern France; to the south, it included land areas along the southern Baltic Sea. Between the

TABLE 1-1.
GEOLOGICAL CALENDAR AND FORMATION OF AMBER.

Era	Period	Epoch	Time	Characterized by	Episodes in the Formation of Baltic Amber	Examples of Various Resinous Flows
C E N O Z O I C	Q U A T E R N A R Y	Recent	Present to 10,000 years ago.	Development of man.	By 3,000 B.C., amber was traded as far away from the Baltic region as central Russia.	Copal, kauri gums.
		Pleisto-cene	10,000 years ago to a million years ago.	Widespread glacial ice.	"Young amber" formed in areas not covered by glaciers.	From deciduous trees in lake regions of Poland.
	T E R T I A R Y	N E O G E N I C	Pliocene	1 to 10 million years ago.	Tertiary Sea invasions; formation of mountains; gradual cooling of the climate.	Amber lumps being transported and redeposited.	Tanzania, Africa resin formed.
			Miocene	10 to 25 million years ago.	Grazing animals developing.	"Blue earth" formed.	From trees producing simetite and Dominican amber resins.
		P A L E O G E N I C	Oligo-cene	25 to 40 million years ago.	Saber-toothed cats.	Drastic changes in topography and climates.	From trees producing rumanite, burmite and Baltic succinite.
			Eocene	40 to 60 million years ago.	Advent of modern mammals.	Growth of the giant conifer forest in pre-Fennoscandia.	Baltic succinite, burmite and rumanite.
			Paleo-cene	60 to 70 million years ago.	Advent of birds.		

Cambrian and Tertiary periods, the limits and boundaries of pre-Fennoscandia underwent many changes.[16]

During Upper Eocene time, gigantic forests yielding amber resins grew on pre-Fennoscandia in a region now submerged beneath the Baltic Sea. The forests were approximately centered at latitude 55° N and longitude 19° to 20° E.[17] However, it is currently accepted that the range of the forest also extended over a greater area, possibly as far south as the Black Sea.[18]

During the Eocene epoch, the climate of pre-Fennoscandia was warm and balmy, its weather ranging from temperate to subtropical. Adequate rainfall encouraged the growth of large tropical trees and a luxuriant vegetation of ferns and mossy ground cover similar to that now found in some areas of South Asia.[19] Numerous now-extinct species of pine, cedar, palm trees, oaks and cypress flourished in this so-called "amber forest." Although paleobotanists are not in agreement as to the exact species of trees from which flowed the resin that formed amber, recent scientific studies indicate the probable existence of several sap-producing species, among which may be 20 to 40 varieties of coniferous trees, all of which produce resins from which amber is the fossilized product. The large number of tree varieties may account for the many variations in color and diaphaneity of amber nodules. It has been found, however, that one tree, thought to be similar to our Arbor vitae or white cedar (*Thuja occidentales*) flourished in greater abundance than other conifers, perhaps outnumbering them ten to one.[20] It is believed that this tree was the major producer of the resins which later formed Baltic amber.

Because more than one tree produced resins (and assuming Baltic amber's source to be several species of conifer trees), paleobotanists coined the name *Pinus succinifera,* to include all such trees. *Pinus* is the Latin for the genus of pine tree, and the Latin word *succus* means "sap" or "gum."[21]

In the past, it was thought climatic changes or a disease must have caused these giant prehistoric pines—some as big as the California redwood—to exude an abundance of sap flow. Some authors were of the opinion that the trees virtually "bled to death" from disease.[22] This view has been discredited by evidence of abundant and apparently unharmful resin flows from trees found growing in New Zealand.[23]

As the resin so produced by the prehistoric trees flowed downward, it occasionally entrapped insects and plant pieces, and eventually accumulated in masses of various sizes and shapes that later became buried in the soil below the trees. Thus, proof of the existence of other trees growing at the same time amber-producing resin was being exuded is found in the remains of leaves, blossoms, twigs, bark, filaments and other similar botanical debris trapped in the amber. Such inclusions have been identified and suggest that relatives of the cypress, cedar and sequoia, as well as other tropical plants now found in subtropical areas of Mexico, the Southwest and southern portions of the United States and the tropical parts of South Asia once grew in the amber forest. It is also believed that the original home of the American giant redwood (*Sequoia gigantea*) was in this immense forest. Other trees of this ancient forest include olives, chestnuts, camellias, magnolias and the cinnamon tree (whose modern representative is now found only in Formosa, Japan and China).[24]

The fossilization process of amber is greatly different from that which converts wood into the stony substance known as petrified wood (whereby the woody cells have been replaced wholly or in part by mineral substances). In contrast, amber retains basically the same organic substances present in the original resin exuded from the prehistoric trees. Extreme conditions, such as pressure by the glacial covering, severe climatic changes and submersion of the resin under salt water, took place over millions of years, causing the process of oxidation. The volatile compounds which imparted "stickiness" to the resin escaped so slowly the resin was prevented from cracking into numerous minute fragments as a result of the shrinkage. During the lengthy time underground, molecules were forced to *polymerize*; that is, to rearrange themselves. This caused a metamorphosis from a tacky resin to a solid, forming a compound with greater stability and hardness than the original substance, and similar in appearance (and in some physical properties) to plastic. (In *no way* is amber identical—chemically or otherwise—to plastic, though plastic often masquerades as amber to the unsuspecting public.)

Modern technology provides the means not only for studying the past through inclusions in amber, but also for isolating and identifying the individual resin components of amber. These are then compared to recent tree resin components to establish genetic relationships between fossil resins and recent resins. For example, the amber found in Chiapas, Mexico has been studied with infrared spectroscopy and is believed to be related to the leguminous tree *Hymenaea,* which produces the recent African resin known as copal, a material sometimes misrepresented as amber.[25]

Pinus succinifera of the Baltic region preserved its own cones, needles and bark in the resin which flowed down its trunk. Thus, it not only provided scientists with a means for identifying and naming the tree, but also for estimating its size. Enormous pieces of amber, weighing as much as 15 kilograms, led paleobotanists to estimate that this stately conifer grew to a height of 100 feet or more.

The variation in places where the sticky exuded resin accumulated resulted in variations in the shape of the consolidated masses or lumps of today's amber. The most common morphological forms of natural amber are the drops (see Fig. 1-11) and stalactites that exuded from prehistoric trees during periods of ordinary resin production. During more abundant exudations, large incrustations in the form of streamlets and lumps resulted. When deposited in crevices or shivered parts of the trunks, amber formed flat, plate-like incrustations, mostly without inclusions and of the purest material.

Over a thousand species of insects and crustacea have been found in amber. Most of the insects are now extinct; they are the ancestors of similar forms existing elsewhere in the world, and they illustrate the high degree of development in the evolution of insects even at this early period. Some species completely disappeared from the Baltic region, but their descendants can be found in warmer parts of the globe.

In time, the climate of pre-Fennoscandia cooled, and movements of the earth's crust resulted in drastic changes in the configurations of land forms. Separation of crustal plates may have resulted in deflection of the Gulf Stream and thus caused climatic changes. In Tertiary time, portions of pre-Fennoscandia sank, older islands vanished and new ones arose from the sea. The ancient amber forest was destroyed, its trees fallen and decayed and the debris either washed away or buried in sands, gravels and clays. These remains formed the clay-rich geological stratum known in the Baltic as the "blue earth." It is this layer that contains most of the amber.

During the ensuing Great Ice Age, the remnants of the continent were covered by vast, moving sheets of ice which irresistibly plowed through the sedimentary formations laid down by erosion of pre-Fennoscandia. The comparatively soft stratum of blue earth was easily pushed away and incor-

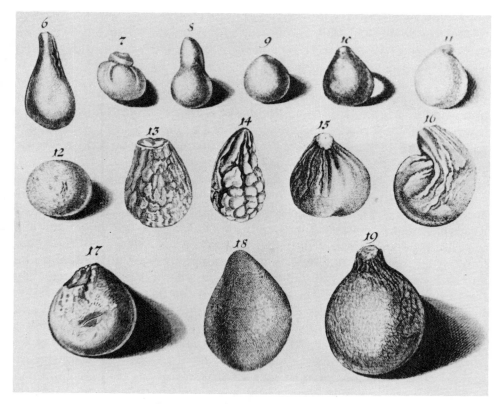

Fig. 1–11. Drawing of natural teardrop-shaped raw amber lumps, from Sendelio in an early (1742) work on amber, "Historia Succinorum."

porated in glacial moraines over much of its extent. It was by this process that Baltic amber, originally confined for the most part of a single stratum, became widely distributed over much of northern Germany and other lands adjacent to the Baltic Sea.[26]

The Great Ice Age occurred during the Pleistocene epoch, approximately a million years ago. The land gradually became free of the icy covering. As water was released, it flowed into depressions carved away by the glacier. Thus the Baltic Sea was formed. Northern Europe had now practically assumed its present forms of vegetation and animal life. The action of the waves on the shoreline cliffs of the Baltic loosened quantities of amber, carried the amber away to sea and deposited it in the sea bed. In the past (and even today), violent storms blowing in the direction of the coast caused—and can still cause—lumps of amber to be thrown out along the shore.

In this simplified overview of geological history of Baltic amber formation, events are described that began 60 to 40 million years ago, in an era just after the age of dinosaurs and extending to the period when mammals began to roam the earth. In terms of geological time, this is a relatively recent development. The vast vegetation of the period that formed amber was the second such period in geological time—the earlier providing the basis for formation of our coal beds.[27]

BALTIC AMBER COAST TODAY

The shore along the northwest side of the Bay of Danzig contains the formation where Baltic amber is found most frequently in the earth layer in which it was originally formed. The drawing shown in Fig. 1-12 is a typical cross-section of the Baltic coast formations which were undisturbed by Ice Age glaciers so that the blue earth was left intact.

Succinite or Baltic amber is found in areas other than the Baltic Sea coast. The southeastern shore of the North Sea along the coastline of Kent, Essex and Suffolk in England produces small quantities of amber. In the past, the western shore of Denmark and the Frisian Islands produced larger amounts than are produced now. Even today, however, after a storm, collectors and local "rock hounds" may walk along Danish shores early in the morning and pick up small pieces of amber. Amber may be found in any area along the German, Polish and Lithuanian shores of the Baltic Sea (as was shown in Fig. 1-5). In spite of this distribution of amber, for the past 100 years, roughly 99 percent of all amber has been produced in the Samland promontory, a high, approximately 20- to 25-mile square ridge of land jutting out into the Baltic Sea. Samland (Kaliningrad Oblast, SSR) was formerly the major amber-producing area of East Prussia.

The cliffs on Samland's coast rise almost vertically from the sea, "like the walls of a fortress," according to naturalist Willy Ley, a native of Samland. In the 1950's, he wrote:

Much of it is forested on top, and the beach at the foot of that strange vertical sandy coast is narrow, so narrow in places that it cannot be walked but has to be waded. If there were mountain goats in the Samland (which is decidedly not the case), they might negotiate these stretches with dry feet by jumping from rock and to rock, because for hundreds of yards out from the coast the sea is dotted with countless 200-ton boulders, brought there originally from the Scandinavian countries by the glaciers of the Pleistocene period and left over when the receding coastline was washed out.[28]

Currently, the largest reserve of Baltic amber deposits is located in the northwest region of Samland, now under the control of the Soviet Union and called the Peninsula of Sambia.[29] Recent reports indicate that the layer of amber-bearing blue earth is buried under approximately 25 to 40 meters of soil, depending upon the exact locality examined. The blue earth varies from 1 to 9 meters in thickness in this region, with the "vein" of amber-bearing earth estimated to be over 2 kilometers in length. Lithuanian geologists believe the second largest reserve of amber is located in the Courish Lagoon, or Kurskiy Zaliv (Russian), a bay previously known as the Kurisches Haff (German) or Kuršiu Mare (Lithuanian). The bay is separated from the Baltic Sea by a long sandbar, which in German, is called the Kur-

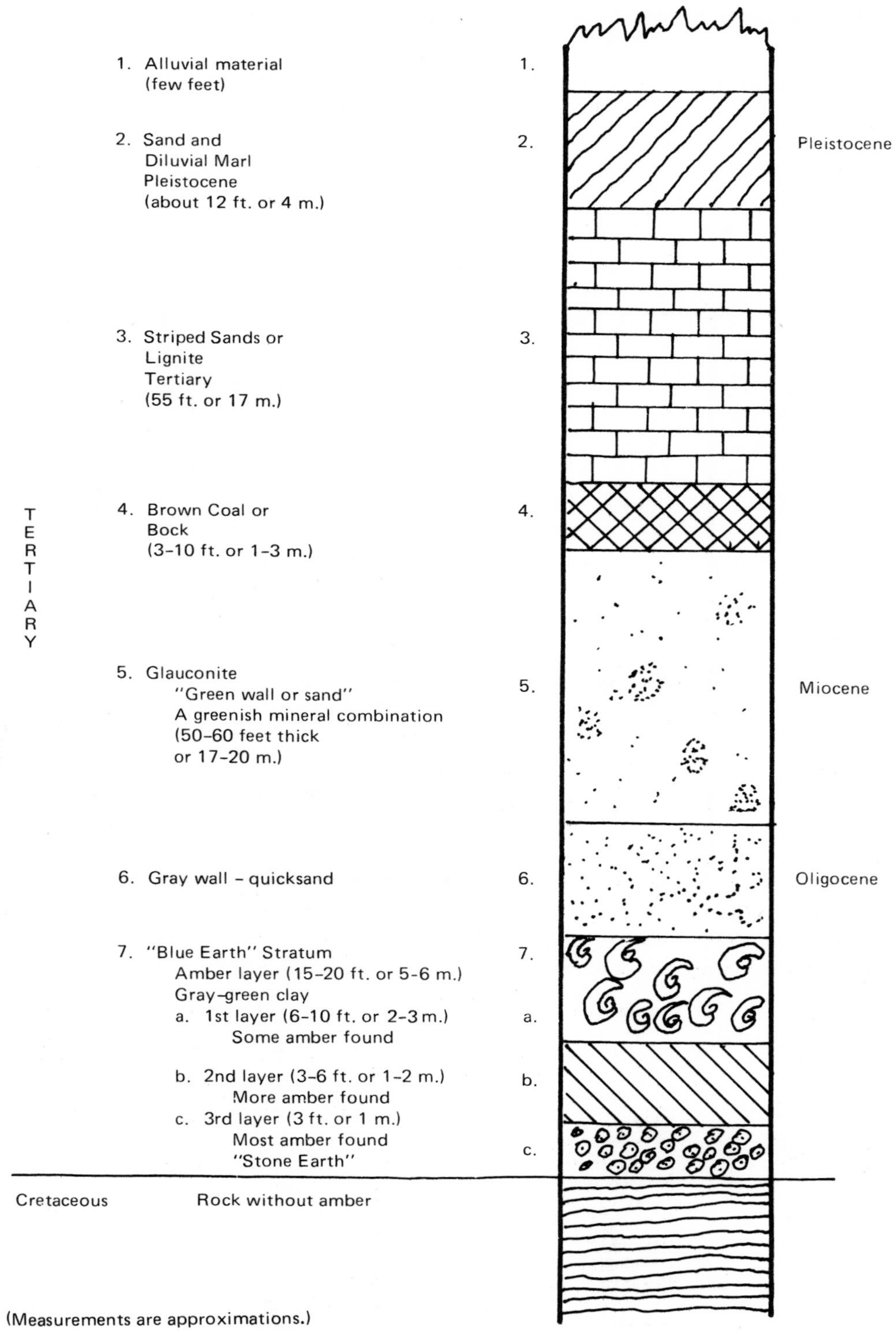

1. Alluvial material
   (few feet)

2. Sand and
   Diluvial Marl
   Pleistocene
   (about 12 ft. or 4 m.)

3. Striped Sands or
   Lignite
   Tertiary
   (55 ft. or 17 m.)

4. Brown Coal or
   Bock
   (3-10 ft. or 1-3 m.)

5. Glauconite
   "Green wall or sand"
   A greenish mineral combination
   (50-60 feet thick
   or 17-20 m.)

6. Gray wall - quicksand

7. "Blue Earth" Stratum
   Amber layer (15-20 ft. or 5-6 m.)
   Gray-green clay
   a. 1st layer (6-10 ft. or 2-3 m.)
      Some amber found

   b. 2nd layer (3-6 ft. or 1-2 m.)
      More amber found
   c. 3rd layer (3 ft. or 1 m.)
      Most amber found
      "Stone Earth"

T
E
R
T
I
A
R
Y

Cretaceous   Rock without amber

1.

2.   Pleistocene

3.

4.

5.   Miocene

6.   Oligocene

7.

a.

b.

c.

(Measurements are approximations.)

Fig. 1-12. Geological cross-section of earth from Samland Region.

21

ische Nehrung; it is now called Kurskiy Merija (Russian) or Kuršiu Nerija (Lithuanian). Amber lumps are occasionally found buried in the sands of the dunes on the Nerija (see Fig. 1–13).

Dredging in the Courish Lagoon has produced vast quantities of amber

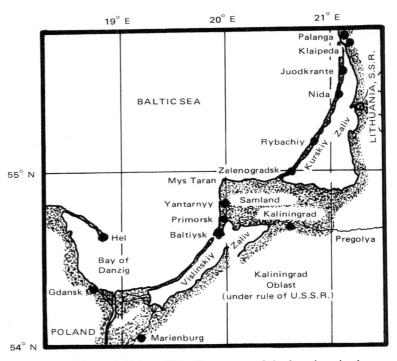

Fig. 1–13. Map of Kaliningrad Oblast, SSR. The names of the locations in the areas where amber is found have changed as the countries have come under rule of various governments. East Prussian names are given below for each location on this map.

Locations on Map	German Name of the Location
Palanga	
Klaipeda	Memel
Juodkrante	Schwarzort
Nida	Nidden
Rybachiy	Rossitten
Zelenogradsk	Cranz
Mys Taran	Brüsterort
Yantarnyy	Palmnicken
Primorsk	Lochstädt
Baltiysk	Pillau
Kaliningrad	Königsberg
Pregolya River	Pregel River
Vislinskiy Zaliv	Frisches Haff
Kurskiy Zaliv	Kurisches Haff
Gdańsk (Poland)	Danzig

since the middle of the ninetheenth century. Geological estimates indicate that approximately 100,000 tons of amber remain in the area.[30]

SUMMARY

The Baltic Sea region has been the original source of amber for the world since prehistoric times, yet amber is not confined to the Baltic area. Even succinite, a product from the Upper Eocene and lower Oligocene epochs, has been found in other parts of Europe and Asia. Amber or fossil resins originating from the earlier Cretaceous period to the Quaternary period are found in various parts of the world.[31]

Although still considered a beautiful and rare gemstone, amber is no longer defined as a mineral as it was in the past. It is now defined as a fossilized tree resin, the word "fossil" indicating a prehistoric origin. Any tree resin becomes "fossilized" through "various changes due to loss of volatile constituents, processes such as oxidation and polymerization, and lengthy burial in the ground."[32] In view of both the ancient uses and the recent understanding of the gem's origin, sparkling amber, which glistens with sun-like rays, can truly be called a "golden gem of the ages."

REFERENCES

1. Williamson, George C., *The Book of Amber*. London: Ernest Benn, 1932.
2. "Amber," *Encyclopaedia Britannica* **I**: 718. Chicago, 1971.
3. "Amber," *Encyclopedia Lituanica* **I**: 84–87. Boston, 1976.
4. Almazjuvelirexport, *Bernstein*. Moscow, 1978, p. 26.
5. *Encyclopaedia Britannica* **I**: 718.
6. Bauer, Max, *Precious Stones* **II**: 536. London: Charles Griffin & Co., 1904 (reprint Dover Publishing, 1968).
7. Helm, Otto, "Mittheilungen über Bernstein 17, Über den Gedanit, Succinit und eine Abart des letztern, den sogenannten mürber Bernstein," *Schrift. d. Naturf. Gesellschaft, N.F. 9,* 1896, pp. 52–57.
8. Bauer, *Precious Stones* **II**: 536.
9. Gemological Institute of America, Assignment No. 33, *Manual for Colored Stones.* California, 1971, p. 2.
10. Bauer, *Precious Stones* **II**: 537.
11. Ley, Willy, *Dragons in Amber*. New York: Viking Press, 1951, p. 37.
12. Bauer, *Precious Stones* **II**: 538.
13. *Ibid.,* **II**: 539.
14. *Ibid.,* **II**: 538.
15. Zalewska, Zofia, *Amber in Poland, A Guide to the Exhibition.* Warsaw: Wydawnictwa Geologiczne, 1974, p. 57.
16. *Ibid.*
17. Conwentz, H., *Monographie der baltischen Bernsteinbäume.* Danzig, 1890.
18. Komarow, W. L., 1943, in Zalewska, *Amber in Poland,* p. 59.
19. Williamson, *The Book of Amber,* pp. 16–17.
20. Sinkankas, John, *Gemstones of North America* **I**: 596. New York: Van Nostrand Reinhold, 1969.
21. Conwentz, H., *Monographie.* 1890.

22. Rath, M., "Golden Amber, The Magnificent Historian," *Lapidary Journal* **25**, *1:* 35, April 1971.
23. Zahl, Paul, "Amber, Golden Window of the Past," *National Geographic* **152**, *3:* 423, September 1977.
24. Williamson, *The Book of Amber,* p. 17. Also in Conwentz, 1890, and E. W. Berry, 1930.
25. Langenheim, Jean H., "Amber, A Botanical Inquiry," *Science* **163**, *3:* 1157–1164, 1969.
26. Williamson, *The Book of Amber,* pp. 19–20.
27. *Ibid.*
28. Ley, *Dragons in Amber,* pp. 4–5.
29. Zalewska, *Amber in Poland,* p. 63.
30. Raimundas, Sidrys, *Third Symposium on the Sciences and Arts.* Chicago: Draugas Lithuanian Newspaper, April 8, 1978, p. 1.
31. Pačlt, J. A., *A System of Caustolites.* Tschermaks mineralogische und petrographische Mitteilungen, dritte Folge, Bd. 3, H. 4 Wien., 1953.
32. Frondel, Judith W., "Amber Facts and Fancies," *Economic Botany* **22:** *4:* 371, October–December 1968.

Part 2
Human Aspects
of
Baltic Amber

2
The History of Baltic Amber

EARLY MAN AND AMBER

It is not exactly known when Baltic amber was first used, but prehistoric amber artifacts found in about 60 localities in Lithuania provide evidence that the ancient ancestors of present day Lithuanians were familiar with amber. It *is* known that early Stone Age man, living near Baltic shores, was aware of the beauty of this material and prized it highly.

Amber's novelty of appearance, its shining luster and its colors made it a treasure so rare that early chieftains took amber with them to their graves, perhaps in the belief that objects buried with them could be used in another world. Such amber hoards provide us with a means for piecing together the story of early barter and the development of trade routes along which amber and other commodities passed from tribe to tribe and from country to country. Northern Europe had other raw materials to trade with peoples in distant southern lands, but amber was one of the most valued. And, fortunately, amber was imperishable enough to leave a record of the trade.

The most ancient recognized period of human development, the Paleolithic Period (or Old Stone Age), is estimated to have lasted over a million years, ending about 8000 B.C. This period is characterized by the use of chipped stone tools. Dates given for Stone Age periods in the Baltic cultures are somewhat later than for other cultures, lasting perhaps as late as 1800 B.C. (Neolithic).[1]

The succeeding Mesolithic Period is estimated to extend from 8000 B.C. to 4000 B.C. According to Edward Strums,[2] a noted Latvian archaeologist, early Baltic people in this period were largely a farming and cattle raising Indo-European tribe, although fierce nomadic warrior tribes who fought with stone battle-axes (and thus earned from anthropologists the label "battle-axe culture") also roamed the area. Eventually, these nomadic tribes intermingled with the agricultural culture, forming a group of people called the "Old Prussians." This group became the forebears of the Western Balts, and they are the same people the Roman writer Tacitas later called the "Aisti."[3]

Some warrior tribes did not join members of the earlier farming culture. These tribes subsequently formed the Stone Age ancestors of the Eastern Balts, today's Latvian and Lithuanian people. The "Aisti" and Eastern Balts picked up amber lumps along the shore of the Baltic Sea and used this resource for their own ornamentation and as a trading commodity.

Conscious of the peculiar beauty and other unique properties of amber, these primitive tribes endowed the material with mystical qualities. Amber discs with designs suggesting a religious meaning (see Fig. 2-1) were discovered in several archaeological excavations. In fact, these are the oldest known symbols indicating the ancient Baltic people were sun-worshippers.[4] The dotted design in the form of a cross radiating away from the perforation in the center was the symbol for the cult of the sun's wheel. The small indentations in the design were filled with resin that provided a decorative accent to the polished surface of the amber disc. The importance of amber in Baltic culture is further indicated by its influence on pottery designs. Balts in the Vistula Basin developed a form of pottery called "face-urns," vessels which bore human features on the covers or necks. Not only were eyes, nose and mouth markings included, but beaded necklaces were depicted around the neck of the urn, and the symbol of the sun on the lid.[5]

The Neolithic Period, also called the New Stone Age, ended about 3000 B.C. Baltic cultures, however, continued in the New Stone Age until 1800 B.C. During this period, characterized by refinements of skill in working with stone, better polished stone implements were manufactured, and the

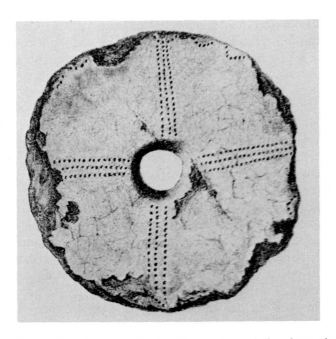

Fig. 2-1. Stone Age amber disc from Sambia Promontory, dating from about 2500 B.C. Discs of this type are the oldest known symbols of sun-worship from the Ancient Balt culture. (From Klebs, *Der Bernstein,* as in Spekke, *Ancient Amber Routes,* 1956.)

oldest amber artifacts found in several places in Lithuania date back to this age.

Amber articles found in Neolithic gravesites along the Baltic Sea coast attest to the local reservation of amber as far back as 5000 years ago. In this period, the region from Sambia and Kuršiu Neriju (Kurishe Nerhung) to Palanga in the north was densely populated. Graves of former inhabitants contain numerous individual pieces of amber jewelry, including pendants, beads, buttons and amulets. Digging for Neolithic artifacts in the swamp regions of the Sventoji River near the Baltic Sea, R. Rimantienė, a Lithuanian archaeologist, found thousands of natural and polished amber articles. However, Juodkrante, a village near Kurskiy Zaliv, in the heart of the amber-bearing region, produced the oldest find and the largest hoard of amber. Juodkrante, before 1940, was called, in German, Schwarzort (East Prussia). It is located on a long sandbar which separates the Courish Lagoon (Kurisches Haff) from the Baltic Sea. By dredging and sifting sand from the bottom of the lagoon in search of raw amber for the industry during the 1850's, searchers found a priceless treasure of amber artifacts. This collection consisted of about 434 Neolithic amber artifacts and was exhibited in the amber collection of the University of Königsberg.

Included among the Juodkrante archaeological findings were five figurines, crudely shaped in stylized human forms (see Fig. 2-2). One portrayed a woman; another, only a face. The collection also included figurines shaped like animals, probably representing a horse or a horse's head.

Amber human figurines were found elsewhere, in numerous Stone Age tombs. Varying in shape and size, they were usually carved from flat pieces of amber and fashioned into small idols or animal shapes. Several of these figurines were drilled through, possibly enabling the owner to wear the piece as an amulet to avert disease and other evil.

The largest portion of the Juodkrante treasure consisted of cylindrical beads, amber buttons, small discs and pendants shaped like stone axes. The amber buttons were lens-like in shape, with cone or V-shaped holes for attaching the buttons to articles of clothing. Such perforations were characteristic of both the Neolithic Age and the succeeding early Bronze Age. Primitive flint and bone drills, employed in shaping and piercing the amber artifacts, left concentric marks in the holes, which provide clues for understanding the past. Apparently, the drills were short and required that the holes be drilled from both sides for complete perforation. When attempting to date these amber findings, archaeologists study such drill markings for indications of technique and to determine the effects of aging on unpolished portions of the amber. But since many surfaces of these primitive amber pendants, buttons and discs are polished, it is often difficult to learn how they were initially shaped.[6]

It was believed that the entire Juodkrante collection was destroyed by fire in World War II, and that these treasures of the past were lost forever.* For-

---

* There is now some doubt concerning the fate of the amber collection. (See the chapter on scientific studies of amber.)

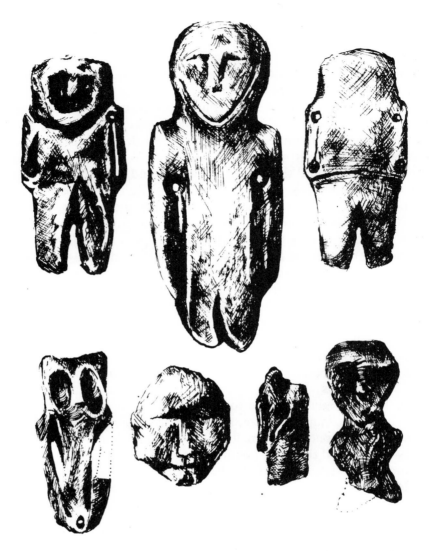

Fig. 2-2. Juodkrante figurines. Stone Age amber amulets from Kurskiy Zaliv near Juodkrante, previously known as Schwarzort. (From Pelka, *Bernstein,* 1920.)

tunately, the collection was photographed before the war, and copies of the original five stylized human figurines were cast in gypsum. Two copies of the figurines were also carved from amber by a Lithuanian amber worker.[7]

Preservation of these links to the "Old Prussian" culture has been a difficult task in the war-torn territory near the Baltic Sea. For example, an amber hoard discovered by a group of amateurs was dispersed among members; thus, a complete study of the artifacts was rendered impossible. One amber figurine shaped into a man-like idol is thought to have found its way into a private collection in New York.

A smaller New Stone Age hoard of amber artifacts was found somewhat farther north, in Palanga, on the coast of Lithuania. It includes 150 artifacts similar to those found in Juodkrante. These are now on exhibit at the Amber Museum of Palanga, along with photocopies and models of the Juodkrante find. The new Amber Museum of Palanga is housed in the mansion of the late Count Tiskevicius, an amber connoisseur himself. By 1963, choice amber collections throughout Lithuania were gathered together and placed in this museum.[8]

The original Palanga collection of amber artifacts belonged to Count Tiskevicius. Some amber articles in the Count's collection were found in peat bogs near Palanga by fishermen searching for raw amber material. Other pieces were discovered in old graves in sandy areas.

The artifacts as a whole represent different periods. Neolithic amber artifacts in the Palanga collection include pendants, beads, discs and buttons with cone-shaped perforations. Some of these are decorated with dots, chiseled lines and pits. Also in the collection is a flat amber carving in human shape, with enough detail to show the waist and a division between the two legs. This unusual piece reveals the skill Stone Age amber carvers developed, despite their crude flint tools. Other articles in the Palanga collection date to the Bronze Age and Iron Ages.[9]

Though amber artifact finds today are rare, the Moscow News Agency, as recently as July 1976, reported that a prehistoric amber hoard has been discovered during excavations near the fishing village of Ventaga in

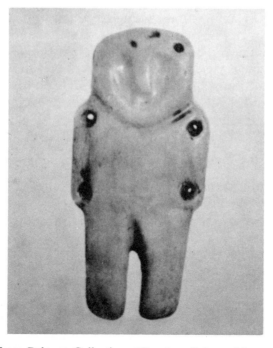

Fig. 2-3. Artifact from Palanga Collection. *(Courtesy Palanga Museum, Lithuania, SSR.)*

Lithuania. The find included 24 pieces, with some amber containing insects, flower petals and leaf imprints.[10]

As indicated by such artifacts, New Stone Age man was superstitious and attributed supernatural powers to amber, perhaps because of its power of attraction: seeing tiny bits of straw and feathers jump to amber lumps after the amber had been rubbed on animal skins must have awed primitive men.

Farther along the Baltic coast, in eastern Latvia, in the region of the Lubana Lake swamps, 2252 amber articles were found by Latvian archaeologist Ilzė Lože. Many of these amber ornaments were not only worn as beads, but were sewn to clothing as decorations and as amulets to impart magical protective powers. Such powers were attributed not only to the shape of the amulet but to the amber as well.[11] Amber buttons were found which were similar to those in the Juodkrante and Palanga collections. These Luban buttons were also perforated by very primitive flint drills, although the shapes of the perforations were different, since V-shaped drills were used rather than the sharp pointed drills used elsewhere in the Baltic region.

The largest Neolithic find, which consisted of a complete Neolithic amber "factory," was discovered north of Liepaja on the Baltic shore near Sarnate by the Latvian archaeologist, L. Vankina, in the early 1960's. The "factory" included flint and bone tools, along with the splinters of flint which were used by the early Balts to cut lumps of amber. The ease with which raw amber can be processed no doubt added to its value for these primitive people. It is no wonder this shining substance was used for amulets and adornment (see Fig. 2-4) by these northern tribes as early as the Stone Ages.[12]

The distribution of amber finds dating from the Neolithic period in the countries of Latvia and Estonia was plotted and mapped by Ilzė Lože. Similarly, R. Rimantienė, a Lithuanian archaeologist, mapped the spread of archaeological diggings yielding amber artifacts in Lithuania. A Latvian archaeologist, Ed. Strums, plotted the Neolithic homesteads and burial grounds in which amber was found in the territory of Germany and Poland. In all, there are now approximately 100 recognized Neolithic burial sites.[13]

Amber Trade During the Stone Ages

Amber was one of the principal commodities for barter in early Europe and the Mediterranean. Archaeologists found amber as far away as central Russia, western Norway and Finland, which indicates the establishment of trade as early as 3000 B.C.[14] Even farther away, some amber ornaments found in Egyptian tombs are believed to date back to the Sixth Dynasty (3200 B.C.) and have been identified as Baltic in origin.

The Aisti and Baltic tribes in the area of northern Poland and Lithuania not only traded their raw amber, but must have traded finished articles as well. Amber beads found in Juodkrante, for example, include those in various stages of completion as well as considerable quantities of finished beads. Furthermore, the finished amber found in Finland, Sweden and Nor-

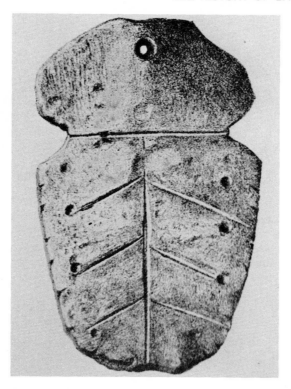

Fig. 2-4. Stone Age amber pendant from Juodkrante, Western Lithuania, dating from about 2500 B.C. Actual size about 3½ inches tall. (From Klebs, *Der Bernstein,* as in Spekke, *Ancient Amber Routes,* 1956.)

way, northwestern Russia and Middle Urals display the same shapes and styles of artifacts found at the source, indicating the pieces were probably worked by the early Balts at the places of origin before being exported.

Amber was also traded into the interior of Germany. An archaeological find near Berlin (Woldenberg, Brandenburg Province) produced artifacts among which was a carving of amber representing a wild horse which dated to the Neolithic[15] (see Fig. 2-5). Amber ornaments were found in remains of Stone Age lake-dwellers of Switzerland and France.[16] In fact, almost every country of Europe has produced amber artifact finds. Such general use of amber during the Stone Age indicates a widely established trade and furnishes evidence of early man's esteem for this beautiful golden substance originating in the northern regions of Europe.

Away from the European continent, the mound tombs (or tumuli) of Great Britain, especially in the vicinity of Stonehenge, are archaeological sites which yielded amber ornaments. The people who built Stonehenge were sun-worshippers and the shiny lumps were thought to be a "substance of the sun." The Great Britain finds range in time from the New Stone Age to the Early Bronze Age.[17]

The last portion of the Neolithic period introduced copper metal tools and, for this reason, it is sometimes designated as the Chalcolithic Period, although it is actually a division of the New Stone Age. Copper was used as

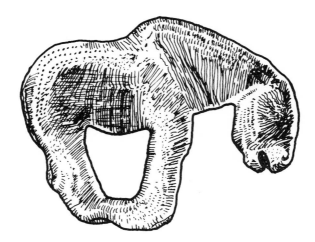

Fig. 2-5. Neolithic amber horse figurine from Woldenberg, Germany. Size about 4 inches from head to tail. (From Pelka, *Bernstein,* 1920; Spekke, *Ancient Amber Routes,* 1956.)

a material for both tools and weapons, but was replaced by the more suitable bronze—a harder, stronger alloy of tin and copper. The Copper Age was of short duration, while the Bronze Age, beginning about 3000 B.C., ended with the advent of the Iron Age (around 1000 B.C.).

Bronze Age and Iron Age Amber Trade Routes

The Aisti trade route of the ancient Bronze Age, by which amber was exchanged for metals, passed along the Baltic coast from Klaipeda to the Elbe River, as shown on the Amber Trade Route Map[18] (see Fig. 2-6). Amber gatherers along the coast enjoyed imported copper and bronze, but cultures farther inland were denied access to metal tools and ornaments and continued to use stone hoes with inserted handles for tilling their soil as late as the Early Iron Age.[19] Old trade routes have now been traced. They provide a record of ancient history, and, as one historian states: "The amber trade route is the original trade route along which luxuries of life went out in search of the necessities."[20]

Some historians believe amber was known to Assyrians even in days of Ninevah (circa 2000 B.C.). A broken obelisk in the form of a tapered four-sided shaft of stone inscribed with Assyrian cuneiform was found in the past century. The cuneiform writing, presumably by a king of Ninevah, was translated in about 1876 by J. Oppert, a noted Assyriologist. Oppert believed the inscription indicated the early existence of commerce between northern Europe and Assyria.[21] His translation follows:

In the sea of changeable winds [indicating the Persian Gulf]
His merchants fished for pearls;
In the sea where the North Star culminates [indicating the Baltic Sea]
They fish for yellow amber.[22]

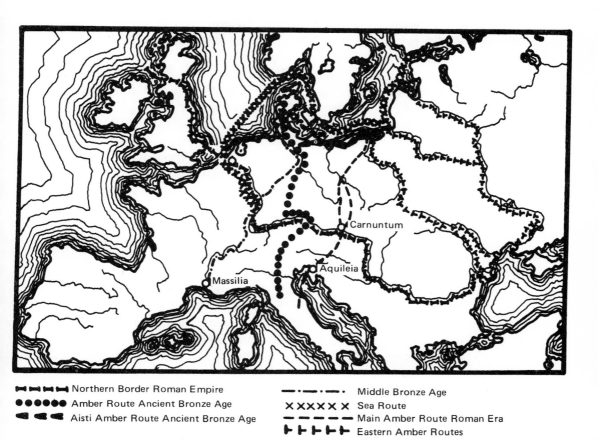

Northern Border Roman Empire
Amber Route Ancient Bronze Age
Aisti Amber Route Ancient Bronze Age
Middle Bronze Age
× × × × × × Sea Route
Main Amber Route Roman Era
Eastern Amber Routes

Fig. 2–6. Amber trade route map. (As in Gudynas and Pinkus, *Palanga Museum,* 1967.)

In 1925, J.M. DeNavarro, the Baltic historian, described early trade routes, as indicated by finds of ancient amber artifacts. The first task DeNavarro set for himself was to determine if this amber was truly succinite, which would then establish it as a product of the north. As he had anticipated, tomb amber proved to be Baltic succinite and not the simetite amber of Sicily, inasmuch as the latter does not contain succinic acid. By plotting locations where Bronze Age tomb amber had been discovered, he delineated the precise routes traveled by early merchants. Latvian archaeologist, Ed. Strums, added to the early work of DeNavarro by the knowledge gained in later discoveries at Juodkrante (Schwarzort).[23]

Research shows that Baltic tribes traded amber mainly along European rivers. During the middle Bronze Age (1800 B.C. to 1200 B.C.), a central route began in Jutland, Denmark (see Fig. 2-6) and passed along the Elbe River southward to what is now Hamburg, Germany. An important hoard of amber artifacts was found at Dieskau, where the route separates with the main branch following the Saale River. From the Saale, amber finds indicate that the route continues south along the Naab to the Danube, thence to the Inn River at Passau on the present border of Germany and Austria. Crossing over the Alps through Brenner Pass, the route follows the Eisock River until it joins the Adige River flowing into Italy.

A second branch of the route followed the Elbe River into Czechoslovakia. Another branch route turned off at Passau and followed the Danube to Linz and Upper Austria. A hoard of amber found at Hallstatt, Austria proved another branch of the Bronze Age amber route followed the Salzach River.

Other amber finds indicate that a western route diverged from the main central route on the Saale, then passed westward just north of the river Main to connect with the Rhine, then traveled south along the Rhine into the Aare River watershed.

Apparently, the Baltic Bronze Age and the transcontinental amber trade for tin and copper began simultaneously. It is known amber was traded to central European cultures, who then traded it to the Mycenaeans and Early Greeks. Amber from the source area of Sambia was shipped to the mouth of the Vistula River, where it was combined with amber from Jutland and then transported southward along the Bronze Age central trading route. Based on the finds of amber in Mycenaean tombs, Marija Gimbutas, a Lithuanian anthropologist, believes this trade began around 1600 B.C.

In pre-Greek cultures, amber played an important role as a luxury and treasure of the educated and ruling classes. Amber beads, pendants and amulets, and hairpins with amber finials, were found in tombs of the Mycenaean kings of Crete. An enormous quantity of flattened spherical amber beads was found among a treasure of golden ornaments in the Mycenaean Acropolis, giving evidence that amber was considered a magnificent substance. While excavating the Acropolis, Heinrich Schliemann found 400 beads which proved to be of Baltic origin. In 1885, Otto Helm,[24] a German scientist, chemically tested Mycenaean amber and identified it as succinite. Using modern infrared photography techniques in 1963, Curt Beck[25] compared the spectra of Mycenaean amber with the spectra of Baltic amber and found them identical, furnishing strong evidence that the Cretan amber was indeed Baltic in origin.

In later periods, amber was exported on a more extensive scale. During the Iron Age, for example, an eastern route beginning at Gdańsk (old Danzig), was established during the early part of the Iron Age, or between 1000 B.C. and 500 B.C. Danzig, in the heart of the Baltic amber region, was a natural starting point for expeditions across Europe to exchange amber for iron, bronze and other commodities. The route followed the Vistula River toward Nakel; then, crossing the lakes between the Vistula and the Warthe (Warta), it continued down the Warthe River into the Oder River near Clagen (Glogow). From the Oder, the route passed to Breslau (Wroclaw, Poland) and over the Carpathian Mountains by way of Glatz Pass, then followed the March River, with the main branch continuing through Carniola and entering the Adriatic Sea near Trieste, Italy. Additional routes branched off to Borno, Czechoslovakia and Vienna, Austria.[26] Still another eastern route began at Klaipeda and followed the Dnieper River, continuing on to the Black Sea region.

The amber trade originating in the eastern Baltic region fluctuated as Mediterranean cultures developed and declined. After the destruction of the Mycenaean Culture by the Central Europeans, for example, the amber supplied to Crete sharply diminished in quantity. The early Balts, however, found new outlets for their amber, which had become known as "Gold of the North," in the Near East.

Near East trade existed as early as 900 B.C., as evidenced by an amber statuette, about 20 centimeters high and representing Assurnasupol, King of Assyria (885 B.C. to 860 B.C.), which was found in archaeological excavations on the banks of the Tigris River. Chemical tests suggest it is of Baltic origin. The exact route amber traveled to reach Assyria is a matter of conjecture, but it may have been traded along routes through Russia and the Caucasus, or through Central Europe and via Phoenician trade routes in the Mediterranean.[27]

The main Eastern Amber Route through Central Europe during the Early Iron Age continued to be used despite invasions by hostile tribes. In the hostilities, the Lusatians of Central Europe were mediators between the Hallstatt Culture in the eastern Alpine area and the amber gatherers. By the beginning of the seventh century, Lusatian intermediaries provided amber to the Etruscans and transported metallic objects from the more highly developed cultures along the amber routes to supply the Old Prussians on the Baltic coast. Such objects included bronze horse-trappings as well as jewelry. The hoards and graves richest in bronze ornaments in the source area for amber were concentrated along the lower Vistula and on the Samland Peninsula. During the period between 800 B.C. and 700 B.C., trade extended into the central Caucasus region. Finds of bronze in new forms, such as belt hooks, ending in two spiral plates, and wheel-shaped pendants, appeared. These are in the style of similar objects found in the Caucasus.[28]

Phoenician Trade in Amber

Sun-worship was widespread, if not universal, during ancient times. Arnold Buffum maintained that Phoenicians brought sun-worship with them from the Orient when they came to the Mediterranean, and that temples built to Baal were temples to the sun. Believing Greeks and Romans to be of the same worship, he stated, "it would be strange, therefore, if amber, which is more sun-like than any gem distilled in Nature's great alembic, should have failed to excite the admiration of sun-worshipping man."[29]

The ancient Phoenicians, a hardy, sea-faring and commercial race, were probably the first sailors to trade amber among the Mediterranean countries, as well as to pioneer sea routes to the Atlantic shores of northern Europe to obtain amber and exchange it for bronze between the thirteenth and sixth centuries B.C. However, amber was probably traded to the Phoenicians through "middlemen" with little knowledge of its place of origin.

In the region between Etruria in Italy and Massilia (the present day city of Marseilles in France) dwelt the Ligurians. Their land, located at the mouth of the Rhone, included the Mediterranean terminal of one of the direct routes from the Baltic. It has been suggested that the Phoenicians obtained amber from the Ligurians, a supposition based on the fact that at one time, amber was called "Ligurian." The Phoenicians were clever traders and went to great lengths to protect their trade secrets. To obscure their sources of foreign goods, they fabricated tales of sea monsters and other dangers encountered on their sea voyages.

Now that the Phoenicians had seen the amber gathered from the sea, they determined to keep the secret for themselves and thus guard the lucrative trade. When the fleets returned to Syria, many were the tales told of perils to the north, of loadstones which would draw the ships to destruction on hidden reefs, of whirlpools which would suck them down to the bottom of the ocean, of witches who enchanted men by turning them into beasts, of terrible sea serpents, and awesome monsters. So well did these ancient sailors spin their yarns that for many centuries afterward mariners feared these mythical perils.[30]

By the time the Roman Empire was established, the earlier Phoenician sources for amber were forgotten. Some Greek and Roman writers reported amber as being dug in Liguria, but Pliny,[31] the Roman naturalist (23 A.D. to 79 A.D.), declared this to be false. He scoffed at Demonstratus, who called amber "lyncurium," believing it to form from the urine of the wild beast known as the lynx—red amber from the male, and white from the female. Demonstratus also informed his countrymen that in Italy there were other wild beasts known as *languri* (thus some people called amber *langurium*). Zenothemis called these wild beasts *langes,* stating that they inhabited the banks of the river Po. Based perhaps on the discovery of insect or plant inclusions, or amber's piney scent when burned, another historian, Sudines, suggested that amber was in reality produced by the *lynx-tree* which grew in Liguria.

Fact and fancy intertwined to explain the mysterious characteristics of this highly desirable golden ornamental substance. Factually established is that with the travels of the Phoenician traders, amber spread to many countries of the Mediterranean, where it sold at the price of gold. As amber was traded for tin in Britain, the Frisian Islands and the Denmark Coast, it would have been within easy access. Thus it was possible for the Phoenicians to trade directly with amber gatherers, eliminating the middlemen as time passed.

The name "Gold of the North," then commonly used for Baltic Amber, is still in use amongst those dealing in amber when referring to the finest amber variety.

Etruscans Trade in Amber

The Etruscans of ancient Italy, another sea-faring culture, and predecessors of the Romans, may have begun trading in amber as early as 1500 B.C. Eventually, they achieved a high degree of civilization and developed extensive commercial relations throughout the Mediterranean region. As evidenced by the artifacts exhumed from their graves and from other archaeological sites, they became extremely skilled in working bronze and gold. Their gold jewelry was especially noteworthy, as they employed techniques such as filigree and granulation, as well as combinations of metal work with gemstones, ivory, glass and amber.[32]

Etruscan women living in northern Italy wore finely polished amber beads. Elaborately designed ornaments discovered in archaeological excavations of this region showed techniques of working amber superior to those displayed in the cruder finished products of the Balts. Although the latter were skilled in perforation and cutting and polishing of beads, discs and simple naturalistic figures in amber, such productions lacked the precise craftsmanship of the Etruscan artists, who were particularly skilled in intricate inlays of ivory and amber.

Amber reached its height of popularity as an artistic medium among the Etruscans during the fifth century B.C., when it was carved into animal shapes and human figurines. Etruscan jewelry includes a wide range of amulets carved into pendants symbolic of fertility. Among them are such motifs as the ram's head, the frog and the cowrie shell. In large necklaces, several of these were combined.[33]

Early Etruscan metal and ceramic objects were commonly encountered in Ligurian archaeological diggings throughout southern France, especially in the Rhone-Rhine areas and in the opposite direction in Greece. These digs indicate important routes for Etruscan trade in amber.[34] It now appears that as early as 500 B.C., the Etruscans learned of Baltic amber sources and established an extensive trade. One hundred years later, Etruscans were probably one of the main importers of amber into the Mediterranean area.[35]

Hallstatt Amber

In the region just north of the Alps, a large hoard of amber was found at Hallstatt in Austria during the 1800's and was dated to the Early Iron Age (see Fig. 2-7). It appears that in the Hallstatt culture, amber was widely used by the peasantry. Amber ornaments in these ancient hoards included hundreds of beads, among which was a necklace containing all forms and sizes of amber beads along with 60 blue and green glass beads. Bronze articles found in Hallstatt represented the entire gamut of Etruscan art, from the earliest Assyrian-Phoenician style to the later Celto-Etruscan mixed forms. Included in the hoard was an unusual type of ornament with decorated plates of bone inscribed with circle designs and associated with large oval beads of amber, indicating a relationship to ancient sun-worship symbols.[36,37]

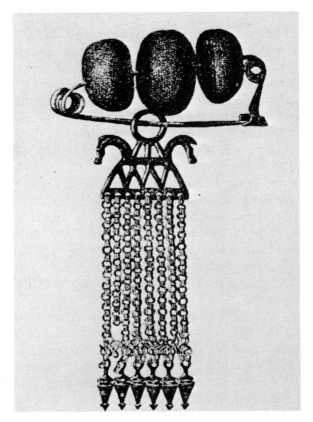

Fig. 2-7. Bronze chain and fibulae with amber beads from Hallstatt (Bronze Age). (From Pelka, *Bernstein,* 1920.)

Greek Cultures and Amber

Despite continual invasion by barbarian tribes, amber remained an important commodity in Greek trade from early Greek cultures onward. Homer, in his *Odyssey* (1000 B.C.), portrays a "cunning trader" enticing female servants to his ship by dangling amber beads before their eyes. In another place, Homer describes echoing halls "gleaming with amber." According to the *Odyssey,* among the jewels offered to the queen of Syria by Phoenician traders was "a golden necklace hung with pieces of amber," And in describing his adventures in Syria, the character Emmaeus remarks, "thither came the Phoenicians, mariners renowned, greedy merchantmen, with all manner of goods, in the black ship . . . there came a man, versed in craft, to my father's house, with a gold chain strung here and there with amber beads, and the maidens in the hall and my lady mother handled the chain and gazed on it, offering him their price."

Among the Greeks, amber was desired mainly for its decorative qualities. Accordingly, it was used in inlays together with ivory. In jewelry, it was used as parts of necklaces and also in pins or *fibulae.* The latter were either made of bronze or gold and were decorated with amber in a manner similar to that illustrated in the Hallstatt object shown in Fig. 2-7. Among the more exotic uses of amber was its use in inlays in harps.

The remarkable property of amber attracting small bits of lint, pith or other light objects after it had been vigorously rubbed was first discovered among the Greeks by the philosopher Thales about 600 B.C. Since the Greek word for amber was *elektron,* it was from this early discovery of the curious attractive properties of amber that our modern word *electron* and its derivatives come. The sun-like color of amber was also of special significance to the Greeks; Homer, for example, described Penelope's necklace as "golden, set with amber, like the radiant sun!" In his descriptions of ancient art objects and wares, Homer indeed mentioned only this gemstone.[38]

The source of amber, however, was shrouded in mystery, and a popular legend said that amber was the hardened teardrops of the Heliades (the sisters of Phaëton) which were shed over a mysterious river called the Eridanus somewhere in the frigid northern vastness of Europe. Herodotus (484? B.C. to 425? B.C.), in his characteristically skeptical manner, not only doubted the existence of any such river but stated he had not found a single person who could say, from his own experience, that there was such a river, or that there was even a fabled sea into which this river was supposed to run.[39]

The actual existence of the Baltic Sea did not appear in historical records until about 46 B.C. Pomponius Mela, a Greek known as "Spaniard Mela," made a first obscure mention of the Baltic. He desribed a bay above the Elbe River full of large and small islands. "In this bay . . . is Scandinavia, inhabited up to now by the Teutons, and it is prominent among the other islands because of its greater size and the fertility of its lands."[40]

Roman Era—"Golden Age of Amber"

For several centuries before and after the birth of Christ, the inhabitants of the Baltic region did not transport the amber along the trade routes to the south, trading it instead to Germanic tribes, their immediate neighbors east of the Vistula River. Trade centers developed in Samland at the mouth of the Nemunas River in the area of the present Klaipeda (Memel) and smaller centers in Galindia and Sudovia, as indicated by the large quantities of Roman coins and artifacts (see Fig. 2-8) found in these areas. The task of maintaining these long and tenuous trade routes across Germany was a constant struggle because of tolls levied by tribes along the routes.

Both Pliny, the Elder (? A.D. to 79 A.D.) and Tacitus (55? A.D. to 120 A.D.), Roman historians, described the influence of amber during the "Golden Age of the Roman Empire." Pliny (actually Gaius Plinius Secundus) wrote the *Natural History,* a 37-volume encyclopedia. In Book 37, entitled, "The Natural History of Precious Stones," he expounded on amber,

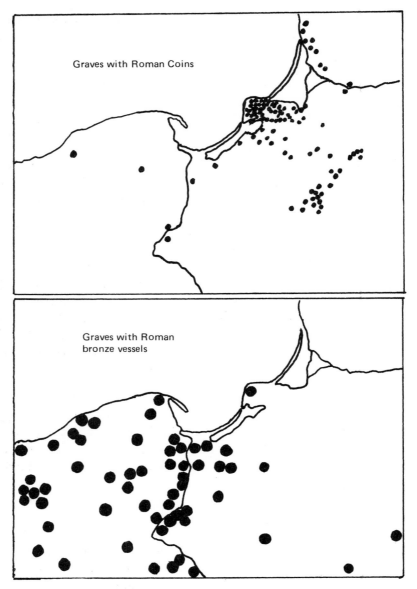

Fig. 2–8. Roman artifacts found in the Baltic Region. (As in Wheeler, *Rome, Beyond the Imperial Frontier,* 1955.)

which he considered to be an object of luxury popular mostly among the women, who had no real justification for its use.

Emperor Nero regarded amber so highly he described his beloved Poppea's hair as having the color of amber. Other ladies of Nero's court dyed their hair to match the color of this gem (and perhaps the color of the hair of the Emperor's favorite!). Since yellow was held in esteem as an imperial color, the demand for amber, as well as its value, increased so much that Pliny tells us "the price of a figurine in amber, however small, exceeded that of a living, healthy slave."[41]

Pliny described several kinds of amber obtained from the "Aisti" amber-gathering tribes:

There are several kinds of amber. The white is the one that has the finest odour, but neither this nor the wax-colored amber is held in very high esteem. The red amber is more highly valued; and still more so, when it is transparent, without presenting too brilliant and igneous an appearance. For amber, to be of a high quality, should present a brightness like that of fire, and not flakes resembling those of flame. The most highly esteemed amber is that known as the "Falernian," from its resemblance to the color of Falernian wine; it is perfectly transparent, and has a softened, transparent brightness. Other kinds, again, are valued for their mellow tints, like the color of boiled honey in appearance. It ought to be known, however, that any color can be imparted to amber that may be desired; it is sometimes stained with kid-suet and root of alkanet; indeed, at the present day, amber is even dyed purple. When a vivifying heat has been imparted to it by rubbing it between the fingers, amber will attract chaff, dry leaves, and thin barks, just in the same way as the magnet attracts iron. Pieces of amber, steeped in oil, burn with a more lasting flame than pith or flax.[42]

At the beginning of Nero's reign, the demand for amber was so great that to obtain a supply for gladiatorial exhibitions, a Roman knight was sent to the north in search of the actual source. Some historians believe this to be one of the most significant historical events of the Roman era because it opened direct trade with Baltic cultures near the Vistula River. The manager of the gladiatorial spectacles, Julianus, commissioned the knight to cross barbarian territories, a feat never before attempted by a Roman.

In attempting to obtain amber from the northern territory for Rome, the knight first had to travel across Roman territory, across the Alps and down the Danube to Carnuntum,* a fortress and the main base for the Roman Danube fleet.[43] (Refer to Fig. 2-6.) Lying about 600 miles north of Carnuntum, the amber coast was eventually reached, but no exact description of the knight's route is recorded. It can be assumed that he followed the amber route along the Marsch River to the Danube, crossing over land to the Vistula River, which empties into the Bay of Danzig. Thus, the existence of a main route for procuring amber for Rome, bypassing middlemen, was established.[44]

Pliny described the knight's influence on the use of amber in Rome after his successful expedition.

This knight traversed both the trade route and the coasts, and brought back so plentiful a supply that the nets used for keeping the beasts away from the parapet of the amphitheater were knotted with pieces of amber. Moreover, the arms, biers and all the equipment used on one day, the display on each day being varied, had amber fittings. The heaviest lump that was brought by the knight to Rome weighed 13 pounds [about 10 pounds U.S. weight].[45]

---

* This famous city was destroyed by the Hungarian invasions during the Middle Ages. Extensive ruins of the city can be found near Hainburg, Germany.

In all, about 13,000 pounds of amber were brought back as a gift to Nero from the German or "Aisti" king. Nero not only adorned the Circus with amber, but made it available to gladiators to wear as amulets or charms on their breasts to assure them victory. Amber pieces were occasionally studded decoratively, like sequins, over the entire garmet of a gladiator. One gladiator's amulet of amber was found with the words "We will conquer" carved on it.[46]

The journey just recounted was probably the cause for shifting the major amber trade from the Elbe River, or central route, to the Vistula, or eastern route. At the same time, the northwestern shore of Denmark and the Frisian Islands diminished in importance in the amber trade, and Samland became the center of production and trade. The extent of Roman trade is shown in the Samland area by discoveries of Roman coins and vases in ancient tombs. Roman coins excavated in the vicinity of the Gulf of Danzig date mainly from 138 A.D. to 180 A.D., suggesting this was the peak period of Roman trade to that region.[47]

In Italy, Aquileia became the main Roman center for manufacture and importation of amber, since it was located at the end of the route to the amber coast. Aquileian workshops manufactured decorative carvings in amber, representing bunches of fruit, animal figures and vegetables. It was customary to give small amber carvings of ears of corn and fruit as New Year's presents. The significance of such gifts is related to the magical properties of amber as well as to the carved forms. Small pots for cosmetics were elaborately decorated in relief, but on occasion large vases also were produced. Amber carvings from the Aquileian workshops are of high quality and are truly exquisite objects of art. So abundant had amber become that the Emperor Elagabalus, in an extravagant display of luxury, paved a portico of his palace with pulverized amber.[48]

The last Roman to write about amber was Tacitus (55 A.D. to 120 A.D.). The eastern trade route, shown earlier as the main source of amber prior to Tacitus' time, had its origin along the western coast of Denmark near the Elbe River and reached the Mediterranean near Marseilles. However, when Tacitus was living, the center of trade had shifted to Samland. In his *Germania,* he described the tideless Baltic as a sea beyond Sweden in an easterly direction "sluggish and almost without movement; which seems to surround the whole of earth because the last rays of the setting sun, until sunrise of the next day, keep the sky so bright that it darkens the stars."

The eastern Baltic sea in midsummer does not darken at night because of its far north latitude. Tacitus was no doubt attempting to describe these "white nights." In another description, he probably refers to what is now familiar as the aurora borealis, which can often be seen from the Samland coast: "Imagination adds that the gods and the flaming coronets on their heads become visible."[49]

Tacitus believed that there were countries and islands in the northern sea where dense forests grew and, stimulated by the sun, produced amber resin in profusion, and that this amber fell into the sea and was washed upon the

German shore. This idea is amazingly close to the truth regarding amber's origin, except for Tacitus' belief that these trees were growing during his own time, when in reality they had long ago been swept into oblivion.

With the weakening of the Roman Empire and its eventual destruction at the hand of invading barbarian tribes, the supplies of Baltic amber to Mediterranean cultures were cut off. All of Europe lapsed into the Dark Ages, and the knowledge of amber was again confined to the region in which it was produced.

AMBERLAND AND ITS ARCHAEOLOGY

While the story of amber has been preserved in the literature of the Mediterranean cultures, virtually nothing except the evidence of exhumed artifacts tells us about the inhabitants of the amber-producing region. What were the early Balts like? How did they regard and use amber?

The answers to these questions are gradually being found by Lithuanian and Latvian archaeologists as they search gravesites dating from 100 A.D. to 1300 A.D. Study of the mounds suggests ancient Lithuanians progressed through three distinct stages of cultural development during this period. A decline of family-oriented systems took place between 100 A.D. and 400 A.D., followed by the evolution of Baltic "class-oriented" communities from 500 A.D. to 800 A.D. and, finally, by the development of an early feudal period (between 800 A.D. and 1200 A.D.).[50]

Numerous amber artifacts found in present day Lithuania along the Baltic seashore in graves and homesteads date to the Iron Age, with their sources shown on the map of Lithuanian archaeological sites. The Lithuanian archaeologists, Gudynas and Pinkus, located 104 graves or homesteads, of which 61 were located in the seashore zone, within 100 kilometers from the sea. A second zone, from 100 kilometers to 200 kilometers from the sea, contains only 33 sites. The most distant zone, over 200 kilometers from the sea, has only 10 sites.[51]

The pattern of these excavations attests to the important role of amber in the lives of Baltic inhabitants. The number of Roman coins and vessels found in the excavations suggests trade with the Romans was most brisk along the shores, where the largest quantities of amber were recovered. It appears that while coastal inhabitants enjoyed an abundance of amber along with the luxuries produced by its trade, amber was scarce even in "Amberland" for peasants living farther inland. Yet it is well attested to by the shape and quantity of amber artifacts that amber was revered by the Balts not only for its beauty but also for its magical properties. The Eastern Balt culture between 100 A.D. and 400 A.D. was rich in superstition and beliefs in the powers inherent in natural objects. For example, spindle whorls, generally found in the "flat-mound" graves of females, were made of amber and thus empowered to prevent evil spirits from "hexing" or tangling the yarn as it was being spun.

Burial mounds of this period in the Shvekshne region contained necklaces

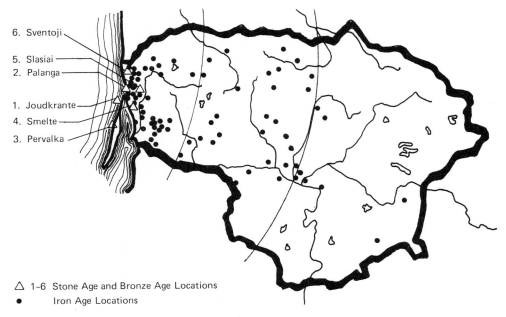

6. Sventoji
5. Slasiai
2. Palanga
1. Joudkrante
4. Smelte
3. Pervalka

△ 1–6 Stone Age and Bronze Age Locations
●   Iron Age Locations

Fig. 2–9. Amber artifact finds from Lithuania. (As in Gudynas and Pinkus, *Palanga Museum,* 1967.)

of amber beads alternated with small blue glass and enamel beads and separated by bronze spirals. Similar amber and glass beaded necklaces dating from this era were excavated in mounds near Palanga (see Fig. 2-10). The glass beads appear to be Roman imports, since the largest quantities are found in sites where many Roman coins were found.[52]

A remarkable concentration of blue and green glass beads was found in Massuria and the Lower Nemunas basin. Presumably, the demand for the bright-colored beads increased as the region developed into a center for fabricating primitive jewelry. However, about the third century A.D., a glass factory began producing not only the blue hemispherical beads, but also the bronze spirals so commonly found in early Baltic jewelry. Some authorities state that amber was so common among the amber-gatherers that they desired something more exotic, such as these glass and bronze ornaments. Samland homestead sites, yielding finished amber jewelry along with pieces still in rough or half-finished condition, prove that amber beads continued to be worked in "Amberland," but, as Marija Gimbutas points out, there was a scarcity of finished amber in relation to the amount of raw material available.[53]

Despite the changing culture from 500 A.D. to 800 A.D., when class systems developed in the Baltic society, the Balts continued to wear amber jewelry. Based on the belief in amber's efficacy as a personal protective material or amulet, amber was given the name "gintaras," meaning

Fig. 2-10. Bronze and Iron Age artifacts from Palanga Museum. Amber beads and spindle whorls, glass beads, bronze spirals and buckles from the first millenium of our era. *(Courtesy Palanga Museum, USSR.)*

"protector." This amuletical power kept amber from being usurped by other gemstones when wealthier persons developed a taste for more elaborate forms of jewelry. Graves from the well-to-do sections of the community contain much jewelry, including silver fibulae, gold plaques and bronze ornaments. A common adornment for women of this period was a collar made from bronze or silver. The construction of these collars portrays the advancement in workmanship and technology of Baltic craftsmen of the era.

New styling is also found in amber jewelry from this period. Bicone amber beads were now combined with characteristic bronze spirals and glass beads. A gravesite in Rùdaičiai (circa 500 A.D. to 800 A.D.) produced a necklace containing 27 small amber beads cut as truncated bicones. A grave of the same period in Lazdininkai contained a necklace made with 13 amber bicone beads, 12 small glass beads and 2 bronze spirals. Flat amber beads were also commonly used.

The style of combining amber beads with small blue glass and bronze spirals was still prevalent during the late Iron Age (or early feudal system period) of the Balts (900 A.D. to 1200 A.D.), although new jewelry materials were coming into vogue. Amber was not allowed to fall into disuse, and strings of small amber beads were widely used because of the persistent belief in the magical powers of the gemstone.

By 1000 A.D. to 1200 A.D., graduated amber beads came into use, with the largest bead in the center and the others diminishing in size toward the ends. Examples of this were found in graves in Nikėlai, Paulaičiai and Švekšna. Bronze collars, first used in an earlier period, were still very much the fashion for neck adornment in the late Iron Age. Some collars were made of a bronze chain with a single small amber bead in the middle.

Surprisingly, graves in eastern regions of Lithuania rarely contain amber articles. Puzzled archaeologists offered several explanations, the most convincing of which related to burial practices. From about 600 A.D. to 1300 A.D., inhabitants living east of the Sventoji River began cremating their deceased; therefore, any amber placed with the dead was completely destroyed. It is assumed that the Eastern Balts used amber to as great an extent as did their amber-gathering neighbors to the west, since, prior to the introduction of cremation, the dead were buried in flat mound graves containing amber articles. Furthermore, raw amber found in alluvial deposits in moraines in Eastern Lithuania, along the banks of the Nemunas River and in numerous other localities, provided readily available raw material. This leaves little doubt that the Eastern Lithuanians used amber widely. Some archaeologists believe amber was prized more highly in Eastern Lithuania than elsewhere and was saved for exchange for imported bronze and iron tools rather than manufactured locally into articles of ornament.

Further insight into the culture of this later Balt period (900 A.D. to 1200 A.D.) is provided by gravesites in the Klaipeda and Palanga region, where more than 300 amber beads were found. Apparently, both women and men adorned themselves with amber, the latter wearing comb-shaped amber pendants on their belts, while the spears of warriors and bridles for horses were decorated with unpolished lumps of amber. The remarkable custom of placing amber in the graves of horses might have been based on a belief in magic. A related custom of braiding one or two amber beads into a horse's mane was probably designed to confer the same protection as the wearing of amber on the person of his owner.

When a woman was buried, round polished amber discs were wrapped in her headdress and placed under her head. Women's graves also contained polished spindles of amber and, occasionally, a small pot containing a natural piece of amber. The significance of these finds, probably owing their origin to the supposedly magical properties of amber, remains unexplained.

Turning to another part of Europe, it is known that from 800 A.D. to 1000 A.D., sea trade was dominated by Vikings sailing unchallenged upon the Baltic. A sea-borne trade to Swedish and Danish colonies thrived, as shown by amber ornaments found in burial mounds upon the Swedish islands of Gotland and Örland. One of the Baltic tribes, the Curonians, inhabiting the region near the Lithuanian and Latvian border, grew powerful

Fig. 2-11. Modern replica of Curonian jewelry. Amber, bronze and hanging chains were typically included in the Curonian style. (Author's collection.)

and began competing with the Vikings for sea trade. The Curonians were a fierce tribe which not only traded amber, but led a war of piracy against the Scandinavian countries, as described by Gimbutas:

Trade and wars of piracy between the Baltic and Scandinavian Vikings continued intermittently throughout the tenth and eleventh centuries. Rich and well-settled Curonia attracted the rapacious Vikings from Sweden, Denmark, and even from Iceland, but they, in turn, were decoyed by the Curonians who plundered their coasts. Thus the powers were balanced by piratical raids on both sides.[54]

The Curonians became the Baltic "Vikings," and they were the most restless and richest of all the Balts during this period.

Curonian graves are rich in bronze and amber ornaments. Their jewelry forms followed geometric patterns with hanging attachments. Several chains were secured to large brooches. Today, this style of jewelry is often copied by Baltic artists (see Fig. 2-11). Curonian ornaments are found all across the Baltic region, attesting to the important role amber played in the life of the forefathers of today's Lithuanians during this little-known era.

REFERENCES

1. Gimbutas, Marija, *The Balts*. New York: Frederick A. Praeger, 1963, p. 19.
2. Strums, Edward, "Der Östbaltische Bernsteinhandel in der Vorchristlichen Zeit," *Jahrbuch des Baltischen Forschungsinstituts, Commentationes Balticae*. Bonn, Germany, 1953, pp. 168–178.
3. Spekke, A., *The Ancient Amber Routes and the Geographical Discovery of the Eastern Baltic*. Stockholm: M. Goppers, 1957, pp. 7–8.
4. *Ibid.*, p. 4.
5. Gimbutas, *The Balts*, p. 65.
6. Gudynas, P. and Pinkus, S., *The Palanga Museum of Amber*. Vilnius: Mintis Books, 1967, pp. 42–43.
7. "Amber," *Encyclopedia Lituanica* I: 84–87. Boston, 1970.
8. *Ibid.*, I: 84.
9. Gudynas and Pinkus, *The Palanga Museum*, p. 44.
10. Hunger, Rosa, *Magic of Amber*. London: N.A.G. Press, 1977, p. 79.
11. Gudynas and Pinkus, *The Palanga Museum*, pp. 42–44.
12. *Ibid.*
13. Spekke, *The Ancient Amber Routes*, p. 5.
14. Childe, Gordon V., *The Dawn of European Civilization, 3rd ed.* New York: Knopf, 1939, p. 200.
15. Ley, Willy, *Dragons in Amber*. New York: The Viking Press, 1951, p. 6.
16. Buffum, Arnold, *The Tears of Heliades or Amber as a Gem*. London: Sampson Low, Marston and Co., 1897, p. 32.
17. *Ibid.*
18. Gudynas and Pinkus, *The Palanga Museum*, p. 36.
19. Gimbutas, *The Balts*, p. 57.
20. Buffum, *The Tears of Heliades*, p. 32.
21. "Journal of Hellenic Studies, XIV (1925)." in Williamson, George C., *The Book of Amber*. London: Ernest Benn, 1932, p. 60.
22. Oppert, J., "L'Amber Jaune Chez les Assyriens (1876)," in Spekke, *The Ancient Amber Routes*.

23. Spekke, *The Ancient Amber Routes,* pp. 48–61.
24. Helm, Otto, *Schriften der Naturforschenden Gesellschaft.* Danzig: N.F. Nr. 2 (1885) pp. 234–239.
25. Beck, Curt, "Archaeological Chemistry," *Science and Archaeology,* Brill, R. (Ed.). Cambridge: M.I.T. Press, 1971, pp. 234-236.
26. Williamson, *The Book of Amber* pp. 61–62; Spekke, *The Ancient Amber Routes,* pp.61–62.
27. *Ibid.*
28. Gimbutas, *The Balts,* p. 83.
29. Buffum, *The Tears of Heliades,* pp. 47–48.
30. McDonald, Lucile Saunders, *Jewels and Gems.* New York: Thomas C. Crowell, 1940, p. 155.
31. Pliny, the Elder, *The Natural History of Pliny* **X**, *37:* 189, Translated by Eichholz, D. E. Cambridge: Harvard University Press, 1962.
32. *Encyclopaedia Britannica, Micropaedia, Ready Reference Index* **I**: 169. Chicago, 1974.
33. Strong, D. E., *Catalogue of the Carved Amber in the Department of Greek and Roman Antiquities.* London: The Trustees of the British Museum, 1966, p. 11.
34. "Etruscans," *Encyclopaedia Britannica* **VIII**: 802. Chicago, 1971.
35. Williamson, *The Book of Amber,* p. 67.
36. *Ibid.,* p. 63.
37. Haddow, J. G., *Amber, All About It.* Liverpool: Cope's Smoke Room Booklets, 1891, p. 5.
38. Buffum, *The Tears of Heliades,* p. 48.
39. Williamson, *The Book of Amber,* p. 30.
40, Spekke, *The Ancient Amber Routes,* p. 77.
41. Pliny, *The Natural History,* **X**, *37:* 201.
42. *Ibid.,* **X**, *37:* Chapter 12.
43. Ley, *Dragons in Amber,* p. 9.
44. *Ibid.*
45. Pliny, *The Natural History,* **X**, *37:* Chapter 11: 199.
46. Szejnert, Malgorzata, *Traffic on the Amber Route.* Poland, source unknown, 1977.
47. Strong, *Catalogue of Carved Amber,* pp. 9–10.
48. *Ibid.,* pp. 11–12.
49. Ley, *Dragons in Amber,* p. 12.
50. Gudynas and Pinkus, *The Palanga Museum,* p. 44.
51. Raimundas, Sidrys, *Third Symposium on the Sciences and Arts.* Chicago: Draugas Lithuanian Newspaper, April 8, 1978.
52. Gudynas and Pinkus, *The Palanga Museum,* p. 44.
53. Gimbutas, *The Balts,* p. 129.
54. *Ibid.,* p. 156.

3
Amber Gatherers

AMBER MONOPOLY AND GUILDS OF THE MIDDLE AGES

After cessation of trade with the Mediterranean because of the fall of the Roman Empire, little was recorded as to the status of Baltic amber and the cultures living along the shores of the Baltic Sea. Haddow[1] reports that on one occasion, during the sixth century, the Aisti, ancient inhabitants of Samland, sent a quantity of amber as a present to Theodoric the Great, the Ostrogoth King. Theodoric not only acknowledged the gift but developed a special interest in amber and sent a mission to distant Baltic shores in quest of the "Gold of the North." Haddow states that Theodoric was fortunate enough to secure a piece weighing from 7 to 8 pounds (3.18 to 3.7 kilograms). During the tenth century, Boleslav, Duke of Poland, succeeded in making Christians out of the pagan Old Prussian Balts, at the point of a sword. In 1161, the Western Balts, or, as they were called, the Borussians, succeeded in casting off Christianity and Polish rule. In the next century, the inroads made by these pagans upon the lands of the neighboring Christians and their advance into Pomerania in 1229 A.D. induced Konrad, Duke of Masovia, to call upon the Knights of the Teutonic Order to conquer these unruly barbarians. Having originated during the Crusades, the knights were proficient warriors and entered the region of East Prussia, where they fought for 50 years, virtually exterminating the Baltic Old Prussians and restoring Christianity to the region.

In earlier times, amber was the absolute property of the finder, but by the time the knights were called upon to conquer the Balts, the Dukes of Pomerania had claimed all amber found for themselves as far as the coast of Danzig (Gdańsk). The control of the amber was the reward given the Teutonic Order for their victory over the conquered Prussians. The amber monopoly, taken over by the Order, was extended along the coast of both West and East Prussia (now Poland and the USSR), with the knights claiming every piece of amber found and any finder being obligated to give up his amber, but receiving very little payment in return. According to Williamson, all the changes in production and sale of amber during the last 800 years were affected by this monopolistic law. Today, newly uncovered raw amber is still regarded as the property of the state in both Poland and Lithuania.[2]

An early document mentioning amber, dated 1264 A.D. and written in Latin, was from the German Grand Marshal of the Knights of the Teutonic Order, granting to the Bishop of Samland the right to fish for amber in a certain locality near Lockstädt.[3] Amber fishing rights on the coast of Danzig were granted in 1312 A.D. to Danzig fishermen, with similar rights to the Monastery of Olivia in 1340, but the only way of receiving such rights was through the Grand Marshal of the Teutonic Order. However, as amber increased in value because it became esteemed for rosary beads, the Order rescinded these treaties, and, by the middle of the fifteenth century, had again gained complete control, earning for themselves the title of "Amber Lords."

In an attempt to organize the amber rosary trade to yield a greater profit, the Grand Marshal issued an order in 1394 forbidding anyone to possess raw amber. All collecting of amber was henceforth to be done under the supervision of a "Beach Master," and all amber was to be delivered to the Amber Lords. This order was brutally enforced. Any unauthorized person observed picking up amber was speedily hanged from the nearest tree. To intimidate local residents, gibbets were erected along the Samland coast, and numerous local legends tell of cruel Beach Masters relentlessly carrying out these

Fig. 3-1. Amber fishermen and gibbets on the amber coast. (From an old copper engraving, Wagner, 1774.)

orders. Legends describe how the ghosts of the Beach Masters were forced in the afterlife to haunt the scenes of their multiple crimes. One such legend tells of ghost riders roaming the shores and inciting the fishermen to revolt against the oppressive amber laws; another portrays one particularly heartless judge, Anselmas of Lozenstein—a man who settled every incident of amber pilferage, no matter how small, with instant hanging—who in the afterworld, as punishment, was condemned to eternally wander the Baltic coast on stormy nights, calling out, "Oh, my god, free amber! Free amber!"[4]

Making sure that every piece of amber was reserved for the Order was only the beginning of organizing the amber rosary trade. After raw amber was collected, it was stored in warehouses in Bruges, Lübeck, Augsburg and Venice to be sorted and worked. The Order prevented establishment of any independent amber works in this own country near the amber source, in the belief that there would be no way to prevent workmen from obtaining amber illegally.[5]

In 1302, amber turners in Bruges were organized into the first of the amber guilds. This was soon followed by the formation of a guild in Lübeck, first mentioned in the Burgher's Register in 1317, with the oldest scroll from the guild dated 1360. Since the Knights shipped amber to guild members to be made into rosaries, the workers called themselves "paternostermakers" or "makers of rosaries." The guilds were assigned their own church for holding divine services and had their own patron saint—St. Adalbert, the

Fig. 3-2. Rosary made of cloudy amber beads from Lithuania. *(Courtesy Gifts International, Chicago.)*

first apostle of the amber region. During this period, the Bruges and Lübeck Corporations produced rosaries for the entire Christian church.[6]

Records of the Order show that in 1399 the Grand Marshal kept his own amber-worker at his castle of Marienburg on the Vistula River in the heart of the amber country. Master Johan, his cutter, made such objects as rosaries, artistic reliefs for altars and mosaics of amber, all of which were entirely for the Order's own needs and not for sale.

Flourishing and growing in power, by 1420, the Bruges Guild was composed of 70 masters and over 300 "helpers and apprentices." (Each master was allowed two apprentices and three journeymen for training.) There was turmoil in the guilds right from the beginning. Fifteenth century records of the Bruges Guild include correspondence with the Teutonic Knights, regarding disputes over raw materials. Appeals were made to the Duke of Burgundy and to the Hanseatic League. Despite disagreements, by the late 1400's, the Pomeranian towns of Kolberg (Kolobrzeg, Poland), Köslin (Koszalin) and Stolp (Slupsk) also had established amber guilds. The actual documents recording the establishment of these guilds no longer exist, but records indicate that the guilds were in existence in 1480. The Kolberg and Köslin Corporations had their headquarters at Stolp.

In 1466, the City of Danzig was taken from the Teutonic Knights and placed under the sovereignty of the King of Poland. The town grew and industry thrived. The Teutonic Order tried in vain to keep amber workers from settling in Danzig. By 1477, despite the efforts of the Order, there were enough craftsmen in the city to form a guild. Because of rivalry with the Teutonic Order, in 1480 the King of Poland granted Danzig the right to maintain its own guild of "paternostermakers."

Although the Order had previously objected to the King's actions, in 1483 the Knights agreed to include Danzig in their trading organization and to supply the Danzig Guild with raw amber. The Order, however, provided competition to Danzig by establishing an amber turner named Hilger, in Königsberg (Kaliningrad). Hilger produced amber products for the Order from 1499 to 1510 and appears to have worked exclusively for the Order. Although the Knights supplied much amber to Hilger, most of it was sent to guilds in Pomerania and Danzig.[7]

During the early part of the sixteenth century, the Teutonic Order, now under the feudal rule of the Polish monarchy, regarded Poland as its enemy. In 1511, the Order, in an attempt to improve its position, appointed Margrave Albert of Ansbach and Bayreuth—a relative of the King of Poland—as Grand Marshal. In 1525, Grand Marshal Albert became a Lutheran and proclaimed himself Duke Albert of Prussia, possessing all the rights of the old Teutonic Order, including the amber monopoly. Continual friction existed between the guilds and the Dukes of Prussia, primarily because the conversion of German principalities to Lutheranism after the Reformation resulted in a lessened demand for rosaries from populations that once were Catholic. With their traditional customers gone, the rosary-producing guilds found it difficult to absorb the raw material which they were obliged to buy. The Danzig amber-turners and several other guilds complained to the German Emperor, requesting mediation between producers and buyers. Production records for this period are meager, but it was reported, in the early 1500's, by Andreas Aurifaber, court physician, that the average yield of amber from the Danzig area was about "110 kegs" with "one year bringing more than another." An old map (dated 1539) that showed the amber sea coast of Samland portrayed the amber packed in kegs or barrels ready for shipping (see Fig. 3-3).

To offset the loss of rosary business, Duke Albert encouraged utilization of amber in new artistic works. He commissioned many large pieces to be carved into beautiful articles. Amber was often used in combination with other gemstones or was combined with ivory and tortoise shell in the making of chests, altar pieces, mirror frames and tables, as well as elaborate carvings of religious figures. In 1681, Duke Albert presented the Tsar of Russia with a "chair of amber." This piece is supposed to be one of the most unusual objects ever made of amber.[8]

In Duke Albert's attempts to again make his amber monopoly profitable, he requested Aurifaber and Göbel, ducal court physicians, to investigate the medicinal uses of amber. Their investigations produced the earliest scientific tracts designed to promote amber as a remedy for sundry illnesses. Both ex-

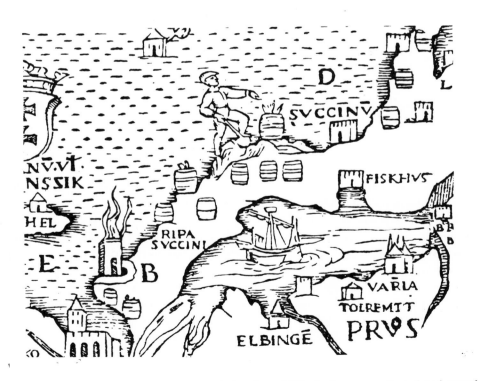

Fig. 3-3. Map of Samland amber coast, dating from 1539. Amber was collected and stored in barrels for shipping. (As in Spekke, *Ancient Amber Routes,* 1956.)

ternal and internal administrations were recommended, especially with white amber as an ingredient (see the chapter on folk medicine).

Finding the sale of raw amber too great a burden, the Duke decided to put the entire trade into the hands of an agent. In 1533, the head of a prominent merchant family in Danzig—Paul Koehn von Jaski—signed an agreement transferring the amber monopoly to his firm, but Duke Albert reserved the right to maintain his own amber-turner to produce amber products for his own needs. White amber was also exempted from this treaty because of the outstanding medicinal properties it was believed to possess.[9] At this time, Duke Albert was annually receiving the equivalent of 15,000 gold dollars for amber, mostly from the Near East trade, since trade with France, Italy and Spain had collapsed. The new monopoly exploited the Near East markets even further by producing prayer beads, a Mohammedan counterpart of the rosary, and thereby created an increased demand for amber. Armenian traders came to the amber country to trade silk carpets ornamented with gold thread for amber goods. Elaborate amber statues, carved by the Danzig Guild for the temples of the Middle East, brought new wealth to the guilds.

Despite new managers for the raw amber industry, fighting continued between the producers and the craftsmen. One well-documented dispute which lasted over a period of years was with the Danzig Guild. Finding it difficult to consume all the raw material they were required to buy despite declining trade, a few master craftsmen complained to the Council of Danzig. On hearing of this, Paul Jaski, though he had recently leased the amber monopoly, refused to supply any amber to the Danzig Guild. At the thought of losing their entire source of livelihood, the guild came back begging for the old terms, with a promise to discipline the masters who had complained. Discovering the guild was at his mercy, Paul Jaski drew up the petition of 1538, in which he agreed to supply the guild with amber as long as he was allowed to sell finished pieces produced by the guild. He then proceeded to give the guild only inferior quality amber. After continued complaints from the Guild and pressure from the Council, a new agreement was made with Jaski around 1546 or 1548. Once again, he promised to sell finished amber, but this time he demanded that the guild's finished goods be sold to him exclusively. Without much foresight, the guild masters promised to "police their own ranks" to guard against craftsmen selling their finished work on their own. The major problem overlooked by the guild was that Jaski had not promised to buy a fixed quantity of goods. Again the guild found itself with more raw material than it could consume.

Complaining to the Council, the guild stated they were "as severely oppressed by Paul Jaski as the children of Israel had been by King Pharaoh." In behalf of the Danzig Guild, the Council requested that Duke Albert sell one-third of his amber directly to the Danzig Guild, bypassing his amber agent. However, placing the troublesome amber monopoly in the hands of his leasee had relieved Duke Albert of many problems. Therefore, wishing to keep peace with Jaski, the Duke ruled in his favor.

The dispute continued. In both 1552 and 1555, the people of Danzig appealed to Duke Albert. Finally, Jaski lost his contract with the amber-turners, but Duke Albert did not sell to them either. This caused problems for the craftsmen, but records do not indicate where or how they acquired their raw materials thereafter. Other sources must have been found, as evidenced by artistic work produced by the guild. Although the House of Jaski controlled the Pomeranian inland amber monopoly, it was limited to East Prussia and the Pomeranian diocese. Other inland areas did not have such strict restrictions on amber as did the Prussian region. Under Polish law, for example, inhabitants of amber-bearing regions could obtain a license to extract amber from their own land, whereas along the shores of Jutland and northern Germany to the mouth of Weichsel, amber could be collected by the owner of the shore. From the Weichsel along the shore near Danzig to Polsk, amber was the property of the City of Danzig. Therefore, the Danzig Guild was able to independently obtain a supply of raw amber. But the principal supply remained in the hands of the amber monopoly controlled by the House of Jaski. (In 1636, the King of Poland, Wladislaus IV, finally commanded Jaski to again sell amber to the Danzig Guild at the same price obtained elsewhere.)

As a sign of the increasing prestige and power of the amber guilds, the Stolp Amber Guild became sufficiently affluent in 1534 to ask for and receive permission from Duke Barnim the Elder to maintain their own brewery. All guild masters were held in high regard, and, in 1555, the Princes of Pomerania referred to the "united, dutiful, time-honored Guild Masters and mutual brothers of the amber-working district of Stolp, Kolberg and Köslin." By 1575, further recognition was given to amber-workers by raising them to the level of merchants. Duke Johannes Friedrich confirmed their rights as trading merchants by edicts dated May 20, 1574 and March 24, 1575.[10] Guild craftsmen gained rights to socially interact and trade with other merchants, receiving the title, "Merchants and Amber Trading Corporation of Stolp." In 1583, the Societies of Kolberg, Stolp, Köslin, Elbing and Danzig united, with Danzig becoming the headquarters of the guild union.

In Königsberg, fine quality amber sculptures were produced by individual artists for European Courts before the formation of the Königsberg amber-worker's guild. It was not until 1641 that the Königsberg Guild was officially formed. As evidence of the skill of its members, it is recorded that a master craftsman of Königsberg, Christian Porschin, invented an amber lens to use as a "burning-glass" to start fires. It was said to be "much quicker in action than were the glass lenses." Having developed a special method to produce transparent amber, Porschin also attempted to make amber "eyeglasses for spectacles." Similar ingenuity was shown by workers in the Danzig Guild, who, upon request, produced a crown carved from a single piece of amber. The crown was presented to John Sobieski, the King of Poland, by the citizens of the city.

One amber-turner of Danzig described by Otto Pelka, in his book on the artistic history of amber, was Michael Redlin. Pelka included a drawing made by Redlin for an elaborate two-tiered jewelry cabinet inlaid with amber and ivory and carved and engraved with landscapes and historical scenes (see Fig. 3-4). Redlin also made a chessboard of multi-colored ambers, complete with matching chessmen. Another masterpiece by this artist was a magnificent 12-branched chandelier which took two years to finish, and which combined amber, ivory and inlaid portraits of Roman and German emperors under clear amber overlays. All three objects were given to the Tsars Ivan and Peter by the Elector of Prussia, Frederick III, but were among the objects lost in Moscow in the destruction of war.[11]

Clear amber was often engraved or incised with decorations on its reverse

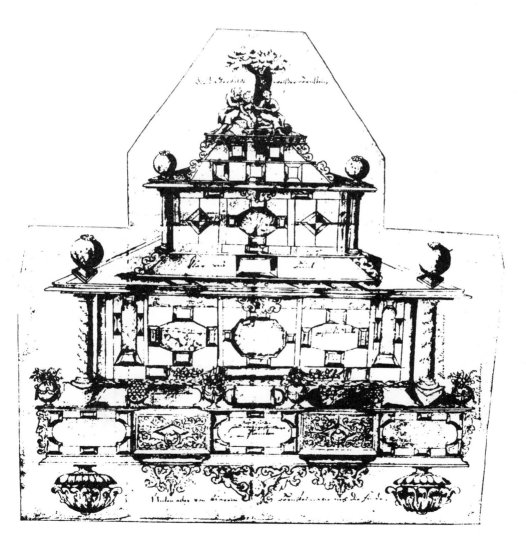

Fig. 3-4. Drawing of a model for constructing an amber jewelry casket by Michael Redlin (seventeenth century). (From Pelka, *Bernstein,* 1920.)

side, a technique adapted from glass-etching technology. In producing coat-of-arms designs, artisans painted parts of the miniature engraving in color, using an art technique similar to that known as verre églomisée. This specialized work was often used on amber in constructing the elaborate pieces of the seventeenth century. Less elaborate forms of this engraving technique on the back of clear amber pieces can be found today. (See Color Plate 19 for samples of modern engraved amber pieces.)

The translucent glowing warmth of amber, often enhanced by placing gold leaf under clear pieces, imparted a bright glow to amber altars designed for small chapels and lent itself well to amber crucifixes, as well as other ritual pieces. Religious altar sets, including chalices, ewers, water sprinklers and candlesticks, were sometimes fashioned of amber. Many of these works of art were destroyed in European wars, but a few examples of these priceless works can still be seen in museums in England, Copenhagen, Stockholm and throughout Germany and Poland.

The amber monopoly remained with the House of Koehn von Jaski until 1642. At this time, Frederick William, the Great Elector, succeeded in retrieving amber rights by paying the enormous sum of 40,000 thalers (approximately $28,000), and thus returned the possession of amber to the state.[12]

New laws were issued to protect the state's interest. Not only was possession of raw amber once again forbidden, but a special permit was needed merely to walk along the beach where amber might be found. The former severity of the amber laws was lessened by making punishment dependent on the quantity of amber stolen, but the punishments were still very harsh: possession of two pounds of amber merited death by hanging, and if more

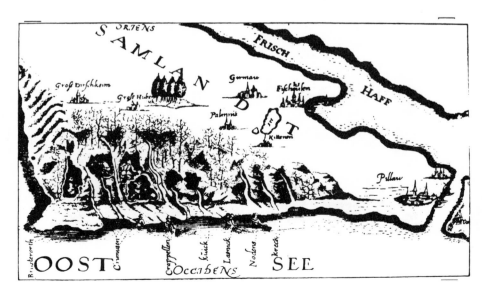

Fig. 3–5. Map of Samland amber coast, dating from 1677, showing amber fishing and mining regions. (As in Stantien and Becker Amber Industry Catalog, Klebs, 1882.)

than two pounds were unlawfully taken or found in possession, the miscreant was put to death by breaking on a wheel.[13] Every third year, fishermen living along the shore had to swear to the "Amber Oath," requiring them to promise to "denounce any smuggler, even if he should be the closest blood relative." The result of this was that the fishing community became riddled with distrust and suspicion, and families were ruined by denunciations, merited or otherwise. An "Amber Court" was a special amber oath for priests who administered the oath and dealt with smugglers further insured that illegal gathering would be held to a minimum. In still another measure, "beach riders" were sent out after storms to prevent local inhabitants from being tempted to pick up the amber which washed up on the beaches. Ironically, the local inhabitants were not only required to turn in all amber found accidentally, but were compelled to go out in search of it. As an incentive, they received salt for the amber, weight for weight, as pay.

Though there was little expense involved in obtaining raw amber, the monopoly was again losing money. Trade diminished steadily. The controllers of the beaches, including beach riders, supervisors, paymasters and amber judges, required salaries. The guilds were becoming impoverished. In 1690, approval for closing the Elbing Guild was requested of the Council, with the members asking that the Council "make binding some privileges of some marriage customs of the masters,"[14] referring to the usual intermarriages between families of guild members designed to exclude strangers. This request was refused by the Council when the guild closed.

The Danzig Guild had difficulties with craftsmen who were not members of the guild, as well as with the suppliers of raw amber. Such persons, working without authority, carved and sold amber goods, and the guild claimed great damage to its members thereby. This damage may have been real, since some of the carvers produced fine quality sculptures and were even commissioned by the Council of Danzig for artistic amber work. One such independent sculptor was Christoph Maucher, who was regarded so highly by the guild in 1705 that they opened their ranks to aid him and gave him instructions for some of his artistic masterpieces.

Not much information is available for sixteenth, seventeenth and eighteenth century guild activities, but artistic works, along with the production of rosaries, continued to provide outlets for the guilds' raw materials. Lübeck Guild records of 1692 describe two elegant crucifixes and a cabinet incrusted with amber made by Johann Segebad and Niklas Steding (who had died the previous year). The "incrustation" process was used to completely cover or "form a crust" over a base of wood with amber. This was actually a veneering process used extensively throughout the period from the seventeenth to the nineteenth centuries. Elaborate chessboards were constructed by this means, with the natural colors and textures of exotic varieties of amber fitted together into mosaics. (See Color Plate 20 for an example of an amber chessboard and carved amber chessman.)

In the early 1700's, the Stolp Corporation became part of the Prussian State, at which time it attempted to gain cheaper rates for the raw amber by

Fig. 3-6. Amber jewel casket. (From Pelka, *Bernstein,* 1920.)

joining with the Königsberg Corporation. On November 3, 1702, the two guilds amalgamated but the union was not always calm, because on at least three occasions, the Königsberg Guild tried to obtain all the raw amber from the Königsberg Chamber of Amber for themselves. Frederick II and Frederick William I upheld the rights of Stolp to share in the amber production.

In 1713 King Frederick William I exhibited the most spectacular of all works of amber art—an entire room, including walls, ceiling and doors, covered with mosaics of amber pieces of varying shades and hues. This magnificent room was furnished with amber vases, dishes, candlesticks, snuff bottles, powder boxes and cutlery, and had taken 12 years and the work of several architects and craftsmen to complete. Eye witnesses reported that "when the sun shone through the windows, it was like standing in an open jewel box." Upon seeing the room, Peter the Great, the Russian Tsar, extravagantly admired its beauty. In 1716, after Tsar Peter's famous victory over the Swedish armies at Poltava, King Frederick William I, in a magnanimous gesture, presented the entire room to the Tsar. The room was carefully dismantled, packed in boxes and taken by sleigh to St. Petersburg, and was reinstalled in the Tsar's winter palace.

In 1755, Tsarina Elizabeth transferred the entire room to the Tsarskoe

Selo palace, adding to it many previous gifts of amber given to the Russian royalty by Frederick the Great, among which was a splendid mirror with the frame carved of amber. The carving depicted the Imperial Russian crown held up by two armed men at the top. The pedestals were carved into representations of the Goddess of War on one side, and, on the other, the Goddess of Peace. Beneath were figures of Neptune and a dolphin, intended to represent Russia's power at sea. At the foot appeared carvings of armor, soldiers and arms, representing Russia's land power. In 1760, several amber carvers from the guilds in Königsberg were called to Russia to complete the carving of the room under the instruction of the Imperial architect, Carlo Rastrelli the Younger.

Curiously, the "Amber Room" vanished from public view in the early 1940's. Though some sources reported that the room was dismantled and hidden for safe-keeping as early as the Russian Revolution, most place a later date on its dissappearance and fear it was destroyed by fire during the siege of 1944. In 1967, the Sunday, March 19 issue of the San Diego Union included an Associated Press news report from Warsaw, Poland, which read:

San Diego Union, B-6 Sunday, March 19, 1967
King's Amber Claimed Buried
Warsaw, Poland AP
Erich Koch, imprisoned former Nazi ruler of East Prussia, was quoted in a Polish newspaper yesterday as saying a valuable "amber chamber" made for King Frederick the Great is buried somewhere in Kaliningrad, Russia, formerly East Prussia's Koenigsberg.

He told the interviewer it is worth the equivalent of $50 million and consists of four wall coverings, including lamps, reliefs, sculptures and frames for mirrors and paintings—all done in amber. It was made at the beginning of the 18th century on the Prussian king's orders and given by his successor to Czar Alexander of Russia.

Another interesting theory related to the room's disappearance is presented in a current book on amber by Rosa Hunger,[15] who reports that the room was dismantled in 1941 and transported to the Prussian Fine Arts Museum in Königsberg. The museum's director, Dr. Alfred Rohde, fearing for the safety of the priceless treasures of art under his care during the bombing attacks by the British, dismantled the room and stored it in an underground cellar. However, when the Russians reached Königsberg in April 1945, the amber room had disappeared. Rohde never disclosed the secret of where the room was stored. No one knows if it was destroyed or whether its hiding place may still someday be discovered.

While the search for the original amber room continues, amber artists from the People's Master Artists of Applied Art in Latvian SSR have taken on the seemingly impossible task of restoring the room. With the help of art critics, librarians, archivists and archaeologists, in 1975, Boris Nikolayevich Blinov, his wife Antonina Georgiyevna and their son Boris collected and thoroughly studied all available records, including those from the former lapidary factory at St. Petersburg, whose artisans restored the amber room at the end of the nineteenth century. To assure that an exact replica was constructed, preserved photos of the lost amber room were enlarged to actual size, making all details down to the smallest piece visible. Colors of the amber were matched and sketches made for patterns so mosaic designs could be accurately shaped. For one panel alone, over 3000 patterns were necessary. To provide an estimate of the amount of amber required for completion of the full-sized room, careful mathematical calculations were made and a model of the room was constructed at one-fifth of its actual size. The model required only 30 kilograms of amber, and thus verification was provided that the contemplated work was feasible.[16]

Despite discouraging opinions of skeptics, the Blinovs completed the first portions of the replica of the original masterpiece: a large amber disc carved with the seal of King Frederick William I, and the upper wreath design of the basic panel. These pieces were presented to specialists of the State Committee for Protection of Cultural and National Monuments in Leningrad and were enthusiastically received. As a testimonial to the work of the artists, the former curator of Kathrine the Great's Palace, and now an Honored Scholar of Culture of the RSFSR,* A. Kuchumov, stated:

> I became familiar with the work of B. Blinov in his restoration of articles of the amber room, and I believe that the principle and general direction of the reconstruction have been correctly undertaken. The fine details of the trim—the carved wreaths, the medallions and other features have been done at a high level of artistry, true to character. The minutest details not only concede nothing to the original, but in some cases surpass the thoroughness of detail. The result can stand on its own as a work of art.[17]

Currently, the Blinovs are continuing the restoration project meticulously, matching every detail of each amber piece on the panels so the completed work will be true to the original made by the guild craftsmen over 250 years ago.

During the eighteenth century, the demand for artistic work in amber began to decline with the development of much bickering among the guilds. By 1755, the Königsberg Guild included only 68 master craftsmen—or the membership the Bruges Guild had attained as early as 1420. The support and interest in amber art by the Elector lessened, and, as the Napoleonic wars developed, it became increasingly difficult to sell amber objects and to exact payment after they were sold, all of which resulted in the decline in commissions for large art objects. In 1811, the Prussian government decided to withdraw from control of the amber industry and again placed the raw amber trade in the hands of a private agent. In the same year, the Königsberg Guild closed and several other guilds dissolved as a result of the collapse of the trade. However, the Stolp Corporation continued making amber necklaces for peasants until 1883.

* Russian Socialist Federation Soviet Republic.

Fig. 3–7. Engraving of amber fishermen on Baltic coast. (As in Treptow, *Bergbau und Hütten wesen*, 1900.)

HISTORICAL METHODS OF GATHERING AMBER

Since prehistoric man picked up the first piece of amber from the Baltic shore, natives of the coastal region of the Baltic have collected pieces washed up from amber-bearing strata which lie submerged in the sea. During storms with strong on-shore winds, large quantities of amber are loosened from the sea floor. Being only slightly less dense than the sea water, they are dislodged easily from the floor deposits and tossed ashore, sometimes becoming entangled with seaweed and flotsam and left high and dry upon the beaches with the ebb of the tide.

Some amber is still found on the beaches and is called "sea-amber", or "sea-stone." Characteristically, it is semi-polished, with only traces of the weathered crust found on buried amber. Furthermore, it is usually free of its fissures because of having been subjected to the violence of the churning sea, as well as to abrasion from tumbling along the bottom of the sea. Thus, masses of amber which originally contained cracks or fissures tend to separate into smaller lumps free from such flaws.

Along the southern coast of Sweden, the western coast of Denmark and the Frisian Islands, native collectors from local villages still gather amber exactly as early man first collected it. After a storm, the beach is full of warmly dressed amber-collectors bucking bitter sea winds as they walk the beach in search of small lumps of amber. Such specimens are generally quite small and they are much scarcer than they were in the past. Amber gathered by combing the beach is sometimes called "strand-amber."

To improve the yield and prevent amber from being washed back to sea, special "amber-catchers," called "Kascher" in German, were constructed of nets fitted to the ends of poles about 20 to 30 feet long. Exactly when this method was introduced is not known, but the first book written on the amber industry, published by Philip Hartmann in 1677, described the procedure which he called "Schöppen" (see Fig. 3-8).

Hartmann pictured an amber fisherman equipped with a long-handled net and clad in protective leather clothing with a pouch or "cuirasse" for stowing amber attached to his chest (see Fig. 3-9). This gear was worn by fishermen who waded waist-deep into the waves, thrusting their long poled nets into the oncoming surf, and scooped up the masses of floating seaweed which perhaps held entangled amber. The contents of the nets were dragged ashore and sorted by the women and children. Amber collected in this manner was referred to as "drawn-amber," or, by the local fishermen, as "scoopstone."[18]

The Baltic Sea was cold in all seasons, but especially so in November and December, when most storms occurred. Gear became frozen in the icy waters and huge bonfires were kept burning to thaw out the cuirasses so they could be used again. The force of the stormy sea often swept fishermen from their feet, and drowning, especially in such clumsy gear, was altogether too common. For protection, some fishermen connected themselves together by ropes, while others dug long poles deep into the sand and attached themselves to the poles when the force of the water pulled them seaward.[19]

During Hartmann's time, from 20 to 30 bushels of amber could be harvested in 3 or 4 hours when amber fishing was most favorable. A bushel of amber weighs about 70 pounds (31.75 kilograms), and even in the late 1800's, when scoop fishing was still a commercial method of obtaining amber, such a rich haul was exceedingly rare.

In marshy areas, the amber search was carried on by men on horseback ("amber-riders") who would ride into the water at low tide, searching for amber left behind by the receding tide. Hartmann described amber divers whose only equipment was a wooden spade carried to loosen the amber from the sea bottom, presumably in warm seasons of the year when the waters were calmer and less cold. Near sand bank areas, the divers descended to the sea floor in the lagoons, which were separated from the Baltic Sea by long sandbars. Using their paddles, they stirred up the sand to dislodge the slightly bouyant amber.

Other methods were devised to collect the material from the floors of the lagoons in the sheltered harbor area from Gdańsk (Danzig) to Klaipeda (Memel). Using long-handled heavy spading forks with prongs bent in a sharp curve (see Fig. 3-10), amber fishermen in broad-beamed rowboats

SUCCINI
PRUSSICI
Phyfica & Civilis
HISTORIA
Cum demonftratione
ex autopfia & intimiori re.
rum experientia deducta
Auctore
M. PHILIPPO JACOBO
𝔥artmann.

FRANCOFURTI
Impenfis MARTINI HALLERVORDI.
Typis JOHANNIS ANDREAE
Anno M DC LXXVII

Fig. 3-8. Title page of the first book published on amber, written by Philip J. Hartmann, printed in 1677.

raked the bottom to dislodge amber lumps which were then scooped up by another fisherman equipped with a small net similar to those used by the open sea amber catchers. This procedure was described in detail by Haddow[20] in 1892:

The men engaged in this put out in boats, each of which has four or five occupants. The work can only be carried on in a clear, calm sea, as the amber has to be fished up from the bottom, and a sharp and practised eye is needed to distinguish it even in the smoothest sea. One boatman loosens the amber with a particular sort of spear, while another holds his kascher, or net, in readiness to catch it. The length of the kascher poles and spear poles varies from 10 to 30 feet. The iron spear head is a plate of iron, the shape of a half-moon, or triangle, three or four inches in length and the same in width. The net is six or eight inches in diameter. When large blocks of stone have to be moved to set the amber free, crooked forks with prongs, sometimes eighteen inches long and twelve inches apart, are employed. During operations, the boat leans over, and the gunwale is brought nearly to the surface of the water.

Fig. 3–9. Plate from the first book on amber (by Hartmann), illustrating amber fishermen in 1677.

In the area of Brüster Ort (now Mys Taran), a somewhat different method was used. Large boulders of the sea floor were hauled up to be used as building stones, and large quantities of amber were often found wedged between the boulders. Amber lumps were raked up with dragnets, the latter fitted with sharp rims to dislodge the small stones and pieces of amber. Hooks were used for loosening the large stones which were then raised to the surface between strong tongs. Amber was obtained in this manner only on the northwest corner of Samland. This tedious method of raking amber from the bottom was called "Bernsteinstechen" or "amber-poking."[21] The methods of "catching" and "poking" for amber may have appeared during

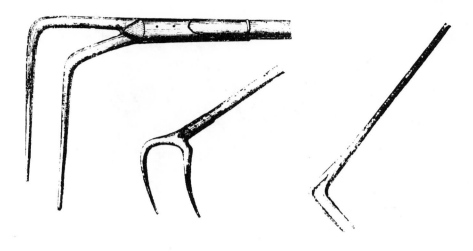

Fig. 3-10. Prongs and forks for dislodging amber lumps from the sea bottom.

the period when the House of Jaski of Danzig (1533 to 1642) controlled the amber monopoly. The increase in supply from these more productive techniques would have enabled Jaski to pay the revenue to the former Grand Marshal of the Teutonic Knights.[22]

In 1725, a new method for obtaining amber from beneath the sea was attempted. The government hired two professional divers from the city of Halle to dive for amber too deep to be reached with forks and probes, but the experiment ended in financial failure. Years later, in 1869, another attempt at diving was more successful, yielding a rich harvest. The firm of Stantien and Becker provided divers with modern equipment to collect amber lying on the sea floor, from Brüster Ort in the west to near Gross-Dirschkeim in the east, until the supply was exhausted.

An interesting account of diving off the amber reef at Brüster Ort describes the costume of the divers:

First a woolen garment covered the diver's entire body. Over this was worn a one piece "India-rubber" suit. The helmet was designed with a small air chest which was connected to an air pump in the boat above by a 40 foot [12.2 meter] long rubber tubing. Another tube attached to a mouthpiece which was held between the diver's teeth. The helmet had three small glass openings to enable the diver to look to the front and to each side. The helmet was screwed on to the diver, a rope was tied around his waist and lead weights attached to his feet, shoulders and helmet. The diver was then ready for his plunge deep into the amber world. For periods up to five hours, the diver walked the sea floor, hooking, dragging, tearing amber from its bed with his heavy two-pronged fork. The diving boats were manned by seven men each—two divers, two pairs of men who worked the air pumps alternately, and an overseer. The overseer stood in the boat to receive amber from the divers' pockets and to see that no pilferage took place.[23] (See Fig. 3-13.)

Fig. 3-11. Amber poking from a boat done on calm days. (From Runge, *Der Bernstein in Östpreussen,* 1868.)

Strict regulations on gathering of amber by the local fishermen were still enforced. In Königsberg in 1826, for example, a permanent executioner was retained to put to death people willfully gathering amber without a lease from the government. In 1811, the government leased the amber rights to the Douglas Consortium for 25 years. However, the enterprise was not successful, and the lease was surrendered before the expiration of the period. After this, local inhabitants were allowed to lease either as individuals or as communities and were then allowed to collect amber by any method they chose. Naturalist Willy Ley[24] describes the bustle of new activity:

For the first time the fishermen worked with real enthusiasm. Travelers who visited East Prussia in that period from 1837 to about 1860 remarked on the very large number of small craft engaged in amber poking; especially of Brüsterort, the "cape" of the Samland, hundreds of vessels could be seen on clear days. For the first time in history, there was no forced labor and no smuggling. And since the beach was now accessible to anybody, the fishing villages could start another type of business; they became resorts. And for the first time in history, the Department of Internal Revenue experienced a succession of years of profit from amber.

Now geologists were able to study the beach areas and lagoons to the north and south of the peninsula of Sambia. Their findings (refer to Fig. 1-14) indicated that the sandbars which protected the lagoons of Frische

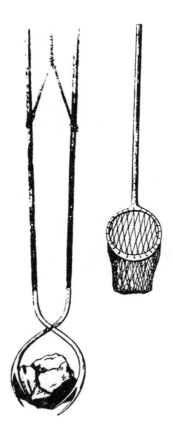

Fig. 3-12. Tongs and net for raising amber from the sea bottom.

Nehrung (near Gdańzk, now called Vislinskiy Merija) and Kurische Nehrung (to the north of Samland, now called Kurskiy Merija; Kuršiy Nerija*) were of relatively recent origin and were still shifting (being blown by the winds) in the direction of the lagoons. Therefore, to keep the Courish lagoon, or Kuriskiy Zaliv,** navigable for shipping goods from Königsberg to Memel, it was necessary to dredge the channel, removing sand blown in from the dunes on the sandbars. Several towns along the Kurische Nehrung were important in this amber shipping route. Goods were brought from Königsberg to Cranzbeek, at the southern end of the lagoon, loaded on barges, and then transported along the inside of the Nehrung, stopped at ports along the way until reaching Memel (Klaipeda). Towards the northern end of the Nehrung was Schwarzort (Juodkrante), where amber amulets and other artifacts from the later Stone Age were found.

The amber stratum in the Kurische Haff is of recent formation, with the amber resting on sea bottom sand among organic debris. At times, amber artifacts similar to those found in Old Prussian (Aisti) graves were found in

* Kurskiy Merija (Russian), Kuršiy Nerija (Lithuanian), Kurische Nehrung (German).
** Kurische Haff (German).

Fig. 3–13. Helmet divers searching for amber off-shore from Brüster Ort about 1869. (After Klebs, as in Treptow, *Bergbau und Hütten wesen,* 1900.)

company with the raw amber. The presence of these artifacts in the Kurische Haff, taken with drawings of old maps, supports the view that the Haff was formerly connected to the Baltic. Repeated inundations of the shores probably washed the artifacts from graves into the Haff.[25]

The possibilities for obtaining rich amber yields by dredging encouraged Whelhelm Stantien, an innkeeper of Memel, to secure a lease from the government in 1854 to produce amber using this method. With only a fishing boat equipped with dredging equipment, he began a successful venture. Near Schwarzort (Juodkrante), in 1855, approximately 1000 pounds (453.59 kilograms) of amber were obtained from the lagoon. Two years later, when dredging was repeated, considerable amounts of amber were again obtained.[26]

In 1869, Stantien obtained financial support by joining merchant Mority Becker, forming the firm of Stantien and Becker. They began work with small manpower dredges, but later employed a steam-powered dredge. In 1861, they were allowed six dredges, paying the state 30 marks per day for dredging privileges. In addition, they were keeping the lagoon free of sand and were required to work at least 30 days from May to September and to pay a minimum of 900 marks per year.

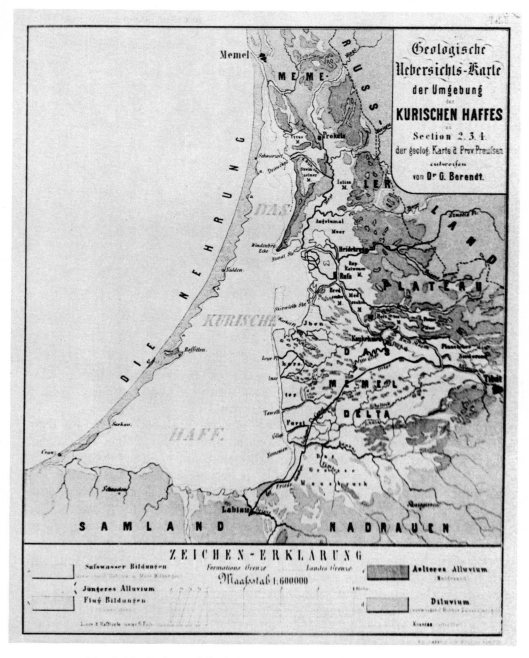

Fig. 3-14. Geology of Kurisches Haff. (Map from Berendt, 1869.)

Six years later, Stantien and Becker had eleven large steam dredges operating. Sand at the bottom of the lagoon was removed to the depth of 35 feet (10.9 meters), dumped on the Nehrung dune and searched for amber. Working time per year, because of severe winters, was about 30 weeks. During the summer, while weather was good, up to a thousand people were

Fig. 3–15. Engraved picture of dunes at Schwarzort, (Juodkrante). (From Berendt, *Geology of Kursches Haff*, 1869.)

employed. In 1868, the dredging operation produced 185,000 pounds (83895.65 kilograms) of amber. On the average, 165,000 pounds (74842.35 kilograms) of amber were produced per year.[27]

In 1882, a new contract was signed, governing amber production from 1882 through 1900, with the payment to the state totaling 200,000 marks per year. The year 1889 produced less amber than did previous dredging, and the following year produced even less. It became apparent the lagoon was becoming depleted. The amber deposits no longer yielded enough to support the dredging expenses. Therefore, in 1890, Stantien and Becker halted operations in the lagoon, but not their search for amber. While dredging had been a success, the firm also experimented in land-mining operations. Previous attempts by the government at obtaining amber from the land simply took the form of raking or scraping, with overseers on horseback guarding the coast and watching for any "suspicious unauthorized digging." Now that geologists were able to examine the Samland coast in detail, the knowledge gained was applied to delineating the geological strata along the seashore. This led to the recognition of the blue earth as the amber-rich stratum. Pit-mining now appeared likely to assure reliable supplies of amber.

Mining attempts had begun as early as 1662 but had not reached the blue earth amber-yielding stratum. Most amber had been found in secondary alluvial deposits, small "nests" of nodules probably washed ashore after storms during earlier geological periods and incorporated in sediments containing silt, clay and sand. Such "nests" were small, sporadically encountered and quickly exhausted. Occurrences of nests were discovered near Palmnicken, in Pomerania and elsewhere along the Baltic coast.[28]

Glacial debris containing amber has been found inland in the Northern Germany lowlands, in Pomerania, Prussia, Lithuania, Poland and Denmark. All of it appears to be derived from the blue earth, which had been excavated by glaciers of the Pleistocene epoch and carried inland. Such unimportant secondary deposits were generally found accidentally during the operations requiring digging, such as excavations for gravel and building. Some of these deposits were mined, but even systematic exploitation tended to be a hit-or-miss venture. Glacial deposits in East and West Prussia, Poland and Lithuania tended to be richest. Near Danzig, at Gluckaw, a glacial deposit was worked for at least 170 years. An unusually large piece of amber, weighing 11 pounds 13 ounces (5.34 kilograms), was found there as late as 1858.[29]

By the end of the eighteenth century, mining was conducted by the Prussian government in the "striped" or "banded" sands of a lignite formation containing some amber just above the blue earth (refer to Fig. 1-12), but was abandoned after 14 years. Such formations are no longer systematically worked because of the expense of the process compared to the yield.[30]

In 1850, a geological survey of the Prussian province of Samland by G. Zaddach[31] contributed much to scientific understanding of amber and its natural distribution. It was he who first stated that all amber originally came from the blue earth layer and not the "striped sands."

Encouraged by information from Zaddach's report, the firm of Stantien and Becker signed a contract with the Prussian government on January 19, 1870, to mine amber. In return, Stantien and Becker paid the equivalent of $1,500 per year per acre of land used. They received exclusive rights to obtain amber by mining, since prior contracts with various communities had expired around 1867 and had not been renewed, as under-financed efforts of mining by the communities had ruined much of the shoreline and destroyed

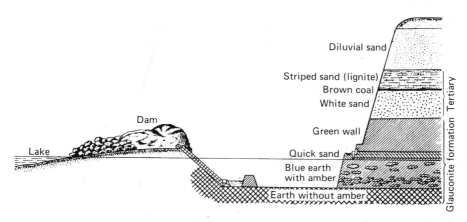

Fig. 3–16. Geological drawing of earth layers of amber-mining area. (From Runge, *Der Bernstein,* 1868.)

Fig. 3–17. Amber mining at Sassau. (From Runge, *Der Bernstein,* 1868.)

otherwise productive land. Near Kraxtepellen, for example, improper techniques of an earlier mining attempt resulted in a long stretch of the coast collapsing into an underground tunnel. As a result of these difficulties, only "shore collecting," "catching," and "poking" contracts were given to communities.[32]

Stantien and Becker's lease allowed them to mine for amber at Palmnicken* between Pillau and Brüster Ort on the west coast of Samland. Meeting with great success, they established a large open-pit operation on the shore at Kraxtepellen. The yield from this mine increased steadily for five years, but because the blue earth amber-bearing stratum in this area was only 6 to 8 meters below sea level, wooden dams were required to keep sea water from entering the diggings. This posed such a problem that in 1879, when the Royal Prussian Mining Administration began working inland from the north coast of Samland at Nortycken, the operation was forced to halt, as water continued to enter the mine pit from the water-bearing sands above the blue earth. Despite drawbacks, Stantien and Becker were more successful in their mining attempts than the government had been. Between 1870 and 1875, they mined about 10,000 pounds (4535.9 kilograms) of amber per year at a yearly expense of 300,000 marks, or at about half the sale value of the amber.

In 1875, Stantien and Becker contracted to build a deeper and larger underground working, the Anna Mine (see Fig. 3-18), near Palmnicken, driving entirely through the blue earth until non-amber-bearing rock was reached. The whole of the amber-bearing stratum could then be removed via shafts, tunnels and galleries, and work did not have to stop during the cold

* Now called Yantarnyy.

season of the year as it did during open-pit operations. A further benefit was that the land above was not destroyed in the process. In the first year, the operation yielded 450,000 pounds (204,115.5 kilograms) of amber, with a similar yield obtained the next year, followed by another 600,000 pounds (272,154 kilograms) in 1877.

An eye-witness description, by Haddow, of the Palmnicken mine portrays the hardships workers endured to obtain amber:

At Palmnicken we visited the diggings in which, about thirty paces from the domain of the waves, the sea-gold is sought. It is an amazing sight! In the downs, shafts and galleries are made. The fresh water is pumped out. Forty feet under the sea level the pits are dug, and the perpendicular boring reaches a depth of fifty feet. The workmen stand in three parallel rows, knocking to pieces every clod of the blue earth, the stratum in which amber is oftenest found. A group of six or eight men is placed under each overseer. While he stands watching, that which is found is thrown into a vessel of water. The men grouped nearest the sea when they have examined the blue earth, throw it with large shovels from the lowest floor of the pit to the higher platform, which is reached by long, narrow ladders. Here the refuse material is taken in charge by a group of men and women, and flung from shovels to the third or uppermost platform, whence it is carted away. All the operations accord with the rhythm of a slow and monotonous melody which the overseers sing. This regularity of movements is intended partly to prevent pilfering which, however, cannot be alto-

Fig. 3-18. The old "Anna" shaft mine workings of the firm Stantien and Becker on the East Sea near Palmnicken, about 1890. (From Andrée, 1924.)

gether prevented, although the miners are carefully searched before leaving the pit after the day's work. It is not astonishing that in the whole range of diggings not less than twenty hundred-weights [100 kilograms] are raised on many a day. Men, women and children in all imaginable costumes, in the oddest of attires, shielding themselves against the sharp, whistling winds, digging vigorously or swinging their shovels to the languid strain of the sombre melody; — what a singular spectacle is this![33]

Though tunnels extended underground, workings were also continued in open pits, where workers were constantly exposed to Baltic winds. Underground mining employed innovative systems for pumping out water, carrying the soil away to be washed, and separating the amber.

By 1885, the mines of Stantien and Becker were producing over 900,000 pounds of amber per year. For the following ten years, the yield fluctuated between 600,000 pounds (272,154 kilograms) and 850,000 pounds (385,551.5 kilograms). Of that figure, 10,000 pounds (4535.9 kilograms) to 12,000 pounds (5443.08 kilograms) per year were produced by shore-gathering and "catching." A record amount of 1,200,000 pounds (54430.8 kilograms) was produced in 1895. Encouraged by this success, the state, in 1899, once again took over the amber mines, with Stantien and Becker receiving a compensation of 9,700,000 German marks (at that time equal to almost $2,500,000).[34] The old Stantien and Becker Anna Mine thus became one of the government mines.

The state established the *Royal Amber Works: Königsberg* to engage in a variety of amber activities in addition to mining. It operated a factory for making amber into jewelry and cigar holders, and chemical plants to produce "amber oil" and amber varnish. The mine at Palmnicken was near exhaustion, but the state continued its operation and opened a new mine nearby. All amber again become state property and had to be delivered to the government works, with cash reimbursement being given by collecting agencies which bought amber from the fishermen. The regulations were so strictly enforced that other firms did not dare to buy directly from the fishermen.[35]

During Stantien and Becker's monopoly, they were threatened by a new invention in 1880 which they worked hard to discredit. It had been discovered that small pieces of amber could be consolidated into large masses by heating to 160 degrees centigrade under high pressure. Given the trade name "ambroid," it was declared an "imitation" by the Royal Amber Works: Königsberg. Nevertheless, the invention was purchased by the state works at a high price. Ambroid was difficult to distinguish from natural amber, and experimentation proved it harder than natural amber and more suitable for smoker's accessories, particularly mouthpieces. Thus the Royal Amber Works also began using small pieces of amber to make ambroid, and from then on, most cigarette holders, cigar holders and pipe stems (see Fig. 3-19) were made of this material.[36]

By the turn of the century, nearly one-half of the total production of

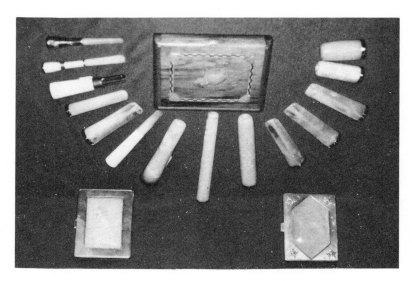

Fig. 3-19. Smoker's supplies made from natural amber, circa 1890's. *(Courtesy Kuhn's Baltic Amber Collection, Florida.)*

amber was devoted to the manufacture of smoking articles. Ornamental objects in a great variety were also manufactured, most commonly round or faceted beads for necklaces, bracelets or rosaries. At the outset, workmen determined the use to which a lump of amber was best adapted to eliminate as much waste as possible. Rough pieces were sorted according to color, form and dimensions in the following classifications:

1. Large, flat and mostly cloudy "tiles" and "plates" for smoking articles.

2. Round, opaque "tears," sometimes flattened at the bottom, for smoking and gem or art articles.

3. Fine and clear "pouches" or shelly amber, which frequently contained insect and plant inclusions, for gem and devotional articles, or for preservation as specimens for museums because of their insect inclusions.[37]

The main finished products were divided into four principal categories:

1. Gems: Necklaces, bracelets, brooches, earrings, pendants, finger rings, cufflinks, teething rings for children, etc.

2. Smoking articles: Cigar and cigarette holders, mouthpieces for pipes, etc.

3. Objects of art: Carvings, jewelry boxes, cups and dishes, writing utensils, ornaments, mosaic pictures, etc.

4. Devotional articles: Catholic, Moslem and Buddist rosaries, sacred figures, amulets, etc.

Such articles were widely exported, with each recipient area having its own preference in regard to the variety of amber used. Beads of the purest, fatty, yellow varieties were most popular with Oriental and English connoisseurs; bone or whitish kinds were preferred by inhabitants of West and East

Africa; light, clear ambers went to the United States; and the finest water-clear specimens were preferred in France. Opaque, inferior varieties were used in Russia and the interior of Africa.

The bulk of the amber was still produced from the government Anna Mine, where it had been opened in 1870 by Stantien and Becker at the foot of a steep 30-meter cliff. The mine now consisted of both open and underground workings. The surface mines were protected against sea invasions by stone dams. Several shafts extended as far down as the blue earth, which occurs in a layer from 13 to 24 feet (4 to 8 meters) thick at this place.[38]

In about 1910, electricity was introduced into the mining works, and its use resulted in increased production. Amber-bearing earth was transported by electrically powered trolleys (see Fig. 3-20) to cube-shaped washing pits, and the cars were tipped into a box holding about 150 cubic meters of amber-bearing soil. Here the amber was sprayed to remove most of the soil. The floating crude amber was then moved to a simple washing plant, where it was washed for ten hours in a revolving drum with clean sand and water to remove dirt and part of the outer crust. The average amber content for the blue earth at Anna Mine reported in 1913 was about 6 kilos per cubic meter.

The cleaned amber was sent to the Königsberg works, which consisted of sales offices, stores, sorting plants and the amber compressing factory. Small grain-size amber lumps (1 to 12 millimeters) underwent further treatment at the state works, rather than going to the hand-workers, and passed through a cleaning apparatus similar in every detail to cereal grain cleaners. The best pieces were then sorted out by hand and either passed on to the sales office or reserved for the compressing factory. The remaining pieces were subjected to dry distillation in the chemical plant to make amber oil, amber acid and amber varnish or colophony.[39]

Fig. 3–20. Transported from the mine in electric rail cars, blue earth was dumped into "wash houses," where the amber was extracted from the soil. (Photo: Amber Works, Königsberg, Prussia, as in Bolsche, 1927.)

No less than 500 women worked at home on cutting and shaping rough amber into jewelry. A weighed quantity of raw amber was given to each woman to be worked in the home by all women of the household. The finished products, together with parings, were returned to the factory, where the materials were again weighed to guarantee all had been returned. Scrapings, dust and chips were sent to the chemical plant for making varnish.

By 1914, hand-working had been largely replaced by machine work. Electric lathes were used to turn amber mouthpieces and the automatic buffing machines used to polish amber pieces destined for jewelry. Girls strung beads for necklaces and made bracelets, with women taking over most finishing operations while men worked in the mines.[40] (See Fig. 3–21)

The state amber factories employed about 1000 male and female workers, in addition to about 500 female home-workers. There were 350 miners permanently engaged in the mines.

In 1911, production at the Anna Mine amounted to a total output of 389,000 tons of blue earth, containing 382,772 kilograms of amber. An additional 15,000 kilograms of amber were contributed by fishermen. The rough amber and by-product sales in 1911 were as follows:

Raw amber	66,700 kilograms
Compressed amber	23,500 kilograms
Colophony	158,200 kilograms
Acid	1,300 kilograms
Oil	31,700 kilograms[41]

As underground mining costs began to increase, and decreasing amounts of amber were found, it became evident that open-pit mining would be more profitable and had to be adopted in the near future, especially since the demand for raw amber was growing and stocks were nearly exhausted. In 1912, production of blue ground was raised 24,150 tons over that of 1911, with a corresponding increase in raw amber amounting to 12,400 kilograms. Production was maintained at 453,592 kilograms per year until the First World War broke out in 1914.[42]

Before World War I, the Lithuanian cities of Memel (Klaipeda) and Palanga also were commercial centers for producing amber. Approximately 500 workers were engaged in the amber industry at Palanga, producing nearly 20,000 kilograms of raw amber per year, mainly from sea fishing and shore-gathering. Similar production was maintained at Memel (Klaipeda), with the addition of amber obtained from dredging.

During the war (1914 to 1918), the Lithuanian amber industry was almost destroyed, but between 1918 and 1933, when Lithuania became independent, it gradually revived. Though no major mines were ever established in Lithuania,* the seacoast of Lithuania adjoining the rich amber beds of Samland furnished an average of about 1 metric ton of amber annually, with

---

* During this period, the Lithuanian industry did not include the mines of the Samland Peninsula, which was then part of East Prussia.

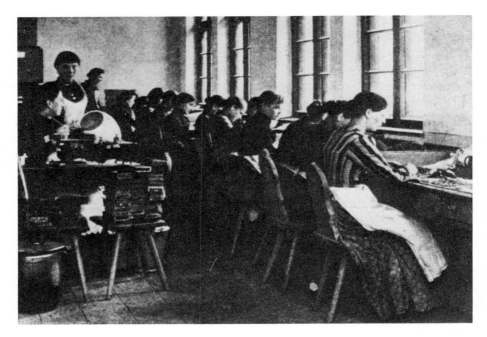

Fig. 3–21. Hand-sorting of amber. (Photo by Gottheil and Son, Königsberg, Prussia, as in Treptow, *Bergbau und Hütten wesen,* 1900.)

the industry controlled by the government. Finders of amber were obligated to deliver the pieces to the state or face a charge of embezzlement. Commerce reports of 1935 indicate about 5 metric tons were imported annually from Prussia to augment the small domestic supply.

Lithuanian amber products were manufactured at ten artisan shops or factories in Palanga, Klaipeda, and Kretinga, in all employing 112 persons. About 5000 kilograms of waste, too small and too dirty for pressed amber, was exported to chemical plants at Palmnicken. One of the products of the Palmnicken chemical plant was amber varnish, which dried to a hard finish but was very dark in color. It was used on ship decks as well as on Stradivarius violins. Besides its dark hue, the expense of the varnish was considered a drawback. Today, partically all varnish is made with synthetic resins, which are more plentiful, more uniform in quality and produce a tougher, longer lasting varnish. Amber acid, or succinic acid, another product extracted from Baltic amber at the chemical plant, was used by iron foundries "as a scum producer for the wet dressings of mineral coal and ores."[43] In the Soviet Republics, amber acid has also been used in the production of soap, bath salts and pharmaceutical preparations, as well as in the manufacture of rhodamine dyes, photographic chemicals and artificial leather.[44] Approximately 15 percent of the products produced by the chemical plants was amber oil, which was used in wood preservatives, insecticides and metal casting and flotation.[45] Amber ore unsuitable even for chemical processing was and still is used in Baltic countries to make an ornamental concrete of greenish and light brown colors.[46]

Preceding World War II, only about 25 to 30 percent of the material produced from the mines operating in the Baltic area was the quality to be used as gem amber and for manufacturing into pressed amber (ambroid). The gem quality material was worked at Königsberg, while the other material was heated in large retorts in the chemical plant at Palmnicken. Today, only one-tenth of the amber production is suitable for jewelry.[47]

After World War I, the Samland amber-producing mines came under the control of the German government. Demand for Baltic amber increased during the 1920's, with amber being second only to diamonds in United States gem imports. Commercial reports show that 15,013 pounds (6809.75 kilograms) of raw amber were imported at a value of $41,556 in 1920 alone.[48] At this time, underground mining met with a number of difficulties which incurred increasing expense. The porosity of the overburden, as well as presence of quicksand just above the amber-bearing stratum, resulted in a change to open-pit mining operations in 1923, and the closing of the old Anna Mine in 1925. By now surface mines were removing approximately 50,000,000 cubic feet (1,400,000 cubic meters) of barren earth overlying the stratum of blue earth per year. Diesel shovels cut away all the overlying deposits down to the amber-bearing blue earth. Other specially designed shovels, using buckets on an endless belt, scooped up the amber-bearing soil and deposited it in conveyors which carried it to the washing sluices.[49] Electric locomotives were now used to carry the blue earth from mines to the

Fig. 3-22. The Anna mine as it looked when it closed in 1925. (Photo by Wilhelm Bolsche, *Bernstein,* 1927.)

Palmnicken factory. Reporting on the status of the mining, Friedrich Prockat described the procedure for recovery of amber:

The cars are emptied over a grate in the so-called spray house, and the blue earth is pulped with sea water under a pressure of six atmospheres, by the flushing methods familiar in the extraction of gold from sands.

The larger amber pieces are collected by hand; the smaller ones flow on with the clay slurry, on which they float because of their low specific gravity—under 1.5. It is thus possible to remove the heavy material from the channel, while the amber-bearing slurry flows on to a classifying drum, where the amber is segregated into different sizes. The larger pieces are cleaned of their crust in special wooden drums to which sand and soda solution are added. At the sorting, which is done at Königsberg, these are again beaten by hand, so that only clean amber will be sent to the market.[50]

Modernization of mining and recover techniques was followed by a record production in 1925 of 1,205,916 pounds (or 547 metric tons) of crude amber.[51]

The Free City of Danzig was also an important manufacturing center for amber products. In 1925, about 600 workers were employed.[52] From 12 to 20 independent manufacturers purchased raw amber supplies from sources not controlled by the Prussian State monopoly, as well as from the monopoly itself, and sold finished amber at their own prices.

In March 1927, a manufacturing and sales agency called *Staatlich Bernstein Manufaktur GMBH* was organized by leading German amber factories, with headquarters in Königsberg. The Prussian state agency, *Preus-*

Fig. 3–23. Open-pit operation for extraction of Samland amber. (From Gottheil & Son, Königsberg, Prussia, as in Treptow, *Bergbau und Hütten wesen,* 1900.)

Fig. 3-24. Blue earth is scooped up by mechanized bucketed conveyers and deposited in rail cars. Open-pit operations in Samland have produced over a million pounds of amber in a single year. (Photo: Staaliche Bernstein-Manufaktur, as in *Natural History,* 1938.)

sische Bergwerks und Hütten-Aktiengesellschaft, was represented in this agency. By 1930, the Prussian State works at Palmnicken employed 600 to 700 men in mining and in the preliminary sorting and treating of amber, with an additional 150 men at the factory at Königsberg. A faceter, generally a woman, was required to work as an apprentice for four years before she was allowed to produce faceted amber beads for the market. About 24 percent of the articles manufactured from amber in Germany were sold in domestic markets. Since Mohammedan rosaries were required to be of amber, substantial markets were established in Mesopotamia and Persia. However, after 1930, production dropped because of unstable commercial and political conditions, and the decline in sales resulted in an accumulation

Fig. 3-25. Modernized wash houses spray jets of water, dissolving the soil so amber pieces float to the surface. Strainers retrieve amber from the muddy sluice. (Photo: Staaliche Bernstein-Manufaktur, *Natural History,* 1938.)

AMBER

STAATLICHE
BERNSTEIN-MANUFAKTUR G.M.B.H.
KÖNIGSBERG i. PR.

Fig. 3-26. Title page of the Staaliche Bernstein-Manufaktur GMBH, dating from 1926 to 1936, showing their trademark. Old faceted beads are still occasionally found in marked boxes or tagged with this trademark.

of stocks sufficient to meet the demand for two or three years. In 1932, production was down to only 59 metric tons.[53]

Because of the political unrest in 1933, when Hitler became Chancellor of Germany, amber production dropped to zero. The National Socialist Party press department attempted to aid the industry and bolster demands for amber by requiring all sporting societies to use amber prizes and decorations for sports awards in place of the usual gold and silver. The resulting improved trade conditions enabled German amber mines to resume operations by March 1934, and 230,000 pounds (104,325.7 kilograms) were produced that year. In 1935, the last year figures were published, 220,000 pounds (99,789.8 kilograms) were produced.[54] During this period, approximately 375 men were employed in the amber industry in Palmnicken.

Amber jewelry has been subject to waves of fashion, being in vogue for a period of two to five years, then followed at about fifteen-year intervals by periods characterized by less demand. During 1934, for example, amber again came into vogue and readily found markets around the world.

By 1939, amber was no longer controlled by the German government, and fishermen of Sambia and Lithuania were free to collect amber as they pleased. This, plus modern methods of production and marketing, resulted in over-production. At the same time, synthetic imitations, produced in large quantities, inevitably led to less demand for genuine amber and a fall in price. (As with diamonds and other precious stones, rarity is one of the chief factors in determining demand and price for objects of luxury.[55]) This over-production, however, was of short duration, and by the mid-twentieth century, supplies of Baltic amber decreased and were made less accessible to the western world as the producing areas were now controlled by the Soviet government.

Few specific facts are known about the amber industry during World War II, but it is known to have been nearly destroyed. Willy Ley reports from personal correspondence that in January 1945, the Königsberg mine was active but the Russians made sporadic air attacks on it. Later in that year, the amber mine flooded and remained so for several years. In 1948, it was reported that the Russians had drained the mine. During the summer of that year they produced 200 pounds (90.7 kilograms) of amber daily. According to Ley:

A newspaper article published in March 1949 in the American Zone of Germany stated that most of the professional amber workers and enough raw amber to last for 20 years were brought to western Germany in 1944 and that the Königsberg Amber Works are now split into two; one in Hamburg (British Zone) and one in Tubingen (American Zone), where amber rosaries (including Mohammedan rosaries) and amber jewelry are being made to satisfy orders from the Benelux nations and from Great Britain. East Prussian amber workers, hearing this, fled through the Iron Curtain. . .[56]

After World War II, amber-processing shops gradually reopened in Palanga, Klaipeda, Kaunas, Plunge and Vilnius in Lithuania. Hundreds of craftsmen were engaged in various phases of the work, processing up to 10,000 kilograms of raw amber annually. Because of political changes in 1963, the amber mines and pits of Samland were incorporated into the Lithuanian amber industry. Lithuanian sources report that these mines produced about 500,000 kilograms of raw amber annually, or about 90 percent of the world's amber production. [57] Currently, two-thirds of the world's supply comes from the concentration of amber-bearing earth in the Kaliningrad Oblast (Samland), but mining in this region is entirely open-pit, with the mine spread over 1200 meters or four-fifths of a mile. The amber is found at a depth of 36 to 40 meters, with the blue earth amber-bearing layer between 5 and 7 meters thick.[58]

BALTIC AMBER CENTERS OF THE USSR

Amber, the Gem of Lithuania

How beautiful our little country,
As the clear drop of amber.
For a long time I love it —
In the patterns of the weavings
And the songs of my native village.

On my palm I bring you
The gentle name of Lithuania
Splendid like the sun, like
A pale piece of amber
Drop of the Baltic —

Salomeja Neris

Artistic works in amber, the "Gem of Lithuania," are treasured throughout the world. However, Lithuania, now incorporated into the Soviet Union, now exports much less amber than it did in the past because less raw material is available. Furthermore, much of the amber is sent to Russian cities and to Moscow shops to supply ornaments for purchase by tourists. The official Russian wholesale outlet controlling the sale of amber, Almaz-juvelirexport, exports natural, pressed and heat-treated amber articles. Amber is so popular in the USSR today that the world's only instructional center for artistic amber-processing is in the School of Applied Arts in Liepaja, Latvia. Graduates of this school work in plants in Latvia and Lithuania, where amber is processed for domestic sale and export. At the present time, only finished amber pieces are exported by the Russian company. West Germany and Japan are the main purchasers of amber today.

Though the largest collection of choice specimens of amber containing inclusions, along with the Juodkrante Neolithic amber artifacts, disappeared from Kaliningrad during World War II, numerous specimens from the amber mines are housed in the Amber Museum of Palanga. This museum, which opened in 1963, presents a unique collection of cultural materials, including the Palanga amber collection, consisting of 100 artifacts similar to

Fig. 3-27. Amber Mosaic illustrating poem about Lithuania. (V. Tomaslunas' collection.)

those in the Juodkrante hoard. Previous to the museum's opening, the Palanga amber artifacts were displayed by Count Tiskevicius in the Paris exposition. In 1936, his collection—which consisted of 153 articles of different ages, from Neolithic to Bronze to early Iron Age—was placed in the Kretinga Museum. It was finally transferred to the Palanga Museum in 1960. In that same year, amber from collections housed in various museums of history, ethnic studies and art, in Kaunas, Klaipeda, Telsiai, Silute, Siouliai, Vilnius and elsewhere were also transferred to Palanga to make it the largest amber museum in the world. The largest and most unusual specimen in the museum weighs 4.28 kilograms and has delicately interwoven streaks of as many as 50 different shades of color. Another interesting exhibit is a mosaic of bits of white and blue amber portraying the atomic icebreaker, *Lenin,* stuck in the ice.[59] Among the exhibits are those explaining the formation of amber, as well as illustrations of the traditional applications of amber throughout the span of Lithuanian history. Unfortunately for tourists from abroad, visitors are allowed only in Vilnius, the capital, and denied access to Palanga and its museum.

In May of 1973, a month-long Baltic Amber Fair, exhibiting the most interesting specimens with inclusions, along with artistic pieces, was held in Paris. Magnificent works of amber, such as necklaces, bracelets, pendants and brooches, made by Baltic craftsmen (particularly from Latvia and Lithuania) were on display, illustrating the ancient Baltic traditions of working with amber. Similar fairs were held in Czechoslovakia and East Germany.

Soviet scientists continue to research the various uses of amber and, interestingly enough, have established that clear amber can be used in laser technology.[60] However, specific details of its use are not given. The more

Fig. 3-28. Lithuanian style constructed with chain and amber chips called "popcorn" amber. (Author's collection.)

conventional use of amber presently in Lithuanian and Latvian SSR is in ornaments for both men and women, the adornment of horses, weapons, vases, umbrella handles and other art objects. Lithuanian artists use early Baltic motifs in their jewelry and combine amber with silver wire and chain. Lumps of semi-polished amber are sometimes strung with silver spirals separating the pieces, and polished amber slabs or beads are spaced with ornamental, twisted silver wires. Very small chips of amber are also strung on slender wires hung to form the so-called "popcorn" clusters, which then are hung upon elaborate chains.

Another popular style of Lithuanian and Latvian artists utilizes natural lumps of amber shaped into large freeform pendants that are commonly set in silver filigree mountings with several silver chains hanging below the pendant. Brooches are often constructed from two lumps of amber connected by links of chain. Such modern styles are based on designs of the ancient Balts and preserve national ethnic styles.

Because people worked so closely with amber, Lithuanian culture, art, customs and mode of life grew together, incorporating the mysterious material from the sea. Local artists, who may have been fishermen, became

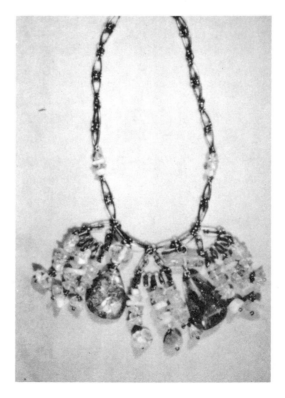

Fig. 3-29. Elaborate chain attachments are typical of Lithuanian jewelry. (S. Kaunelis' collection.)

Fig. 3–30. Silver twists separate each amber bead following ancient Baltic style of construction. (Author's collection.)

Fig. 3–31. Amber brooches made in Lithuania with chain attachments. (M. Kizis' collection.)

Fig. 3-32. Amber brooch with an insect inclusion and designed after ancient Balt styles. (Author's collection.)

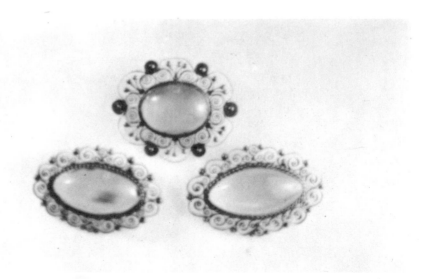

Fig. 3-33. Brooches with amber and Lithuanian filagree metal work. (H. Evens' collection.)

masters in working amber. Carving amber lumps to pass away the dreary winter months, they produced numerous ornaments which are called, in Lithuania, "Kniepkiniai." Because of their strong love of nature, local artisans felt that amber was more beautiful when the least shaping took place, and every effort was made to retain the natural shape. The school of Lithuanian Art emphasizes simplicity of design and the inherent beauty of amber as seen in ancient amulets and other primitive ornaments. Thus, large lumps of amber are often given just enough polish to enhance their natural beauty and then simply mounted on wooden bases, as shown in Fig. 3-34, where an unusually large lump weighing over 1 kilogram has been treated in this manner.

Not all work is so simple, however, and animal figures and other objects from nature are common subjects for carvings and sculptures in amber. A typical sculpture depicts clusters of growing mushrooms, a great delicacy among the Lithuanian people since an abundance grows in wooded areas of the Lithuanian countryside. (See Color Plate 24.) Mushrooms have long been regarded as a symbol of fertility throughout Europe.

Devotional objects, such as rosaries, medallions, crucifixes and religious art, are also made of amber, but these are becoming difficult to find and are mainly for export, since the carvers are discouraged from producing them in

Fig. 3-34. Amber lump weighing over 1 kilogram, from Lithuania.

Fig. 3-35. Mushroom sculpture made of natural amber.

the officially atheistic culture of the Soviet Union and its dependencies. Older amber mosaics depict local myths and religious monuments, such as the "Chapel of Lourdes," which is located on a hill in Palanga (see Fig-3.36) or the Miraculous Mother of the Gate of Dawn (see Fig. 3-37), found in a

Fig. 3-36. Chapel of Lourdes depiction in natural amber mosaic from Lithuania, (V. Tomaslunas' collection.)

Fig. 3-37. Amber mosaic of scenes from Vilnius, the capital of Lithuania. Miraculous Mother of the Gate of Dawn. (V. Tomaslunas' collection.)

chapel in Vilnius. Lithuanians would go to the Gate of the Dawn Chapel to place symbols representing their illnesses or their problems, believing the miracle-working Holy Mother would then cure them.

During festivals, golden strands of amber, the national gem, and other amber jewelry are still traditionally worn with the official national costumes. In the United States, objects such as sculptures, mosaic scenes and jewelry constructed of amber can still be purchased in the Lithuanian community in Chicago, where many shops have sculptures combining brass and amber crafted by a local artist, as well as amber goods imported from Lithuania.

AMBER IN POLAND

During the fifteenth to the eighteenth centuries, masterpieces of sculptural art in amber were produced at craft centers in Gdańsk (Danzig), Slupsk (Stolp) and Elblag (Elbing), all located in what had been East Prussia before the war. After 1940, the Free City of Danzig and portions of East Prussia were added to Poland and city names changed as shown.

Amber sculptures and altar pieces in Polish churches were mostly

Fig. 3-38, Fig. 3-39. Bronze and amber sculptures made by Lithuanian artist in Chicago. *(Courtesy Gifts International, Chicago.)*

destroyed during World War II, but today several museums in Poland contain historical exhibits of amber. The archaeological museums of Lodz and Gdańsk display collections of medieval amber products, with detailed descriptions for each period. The region of Kurpie, Poland, where a cottage industry in amber has existed since medieval times, is represented by a collection of hand-made articles in the Kurpiowskie Museum of Lomze. A collection of artistically sculptured amber pieces from the sixteenth and seventeenth centuries, as well as modern carvings, is on exhibit in the Castle Museum in Malbork. The largest collection, however, is in the Museum of the Earth in Warsaw, which not only explains amber's formation, but also the history of its use from antiquity to the present.

Supplies of natural amber are scarce in Poland and the industry is regulated by the state. Therefore, the collection of raw amber is controlled by a committee of the National Council of Gdańsk, which grants licenses to individuals to mine amber near Gdańsk. Amber is not mined on as large a scale as it was before World War II because many deposits productive in the past are now exhausted. Instead of underground excavations, amber is washed out of the earth by using large, high-pressure water hoses to inject water into the amber-bearing layers and thus force the amber lumps to the surface. Unfortunately, this method disrupts and despoils the ground, and much controversy exists between environmentalists and the amber authorities. The map of Poland in Fig. 3–40 shows regions where amber is generally found.

In the vicinity of Gdańsk, some small firms and cooperatives wash layers of sand to recover amber. One small firm outside Gdańsk turns over about 8 tons of raw amber per year by using a workforce of 15 miners and processors. Such a firm may supplement its own production by importing raw amber from Kaliningrad on the Soviet side of the border. Besides the private firms producing amber goods, Gdańsk is also the location of the State Amber Works, or *Pánstwawa Wytwórnia Wyrobów Bursztyowych*. Natural amber jewelry items, including Moslem rosaries, are produced here and exported to Mohammedan countries in the Near East. Pressed amber necklaces and large statuettes of artificial amber are also produced for export to countries throughout the world. Elsewhere, a mechanized workshop for processing amber has been in operation since 1939, at Slupsk, in the location of the old Guild of Stolp, which closed its doors in 1883. In Sopot, the

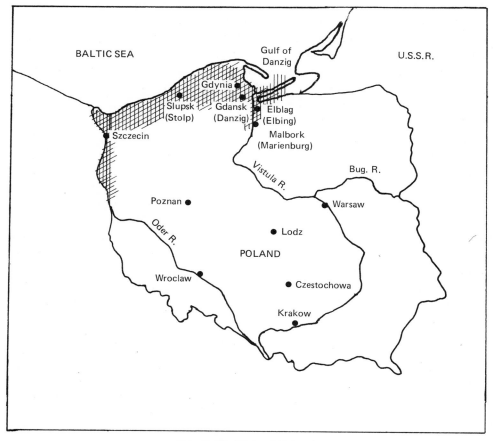

Fig. 3–40. Map of Poland.

wholesale amber firm, the Art-Region Cooperative, employs artists and craftsmen in a factory and also farms out work to many home piece-workers. This firm also has a factory at Wryeszez, a suburb of Gdańsk.

An artist working at the Orno Cooperative in Warsaw describes his feelings about amber thusly:

> To me, amber seems closer to human nature than any precious stone. It is warm in color and to the touch, gay, luminous, and full of surprises . . . It gives me pleasure to find remains of ancient life when polishing a piece of amber . . . What can be more beautiful than a lump of sunshine given the minimum of polish not to spoil its natural charm and set in silver which contrasts so well with its color and substance?[61]

Amber imported from Poland is set in silver* or a silver-colored alloy metal, since Poland does not allow gold to be taken from the country. Craftsmen of the amber-working cooperatives near the Baltic all produce essentially similar products for export. The articles are channelled through Cepelia, the official government chain of shops throughout Poland. All direct export from Poland of amber jewelry must be done by Cepelia, and correspondence from the Polish government states, "Coopexim-Cepelia is the sole exporter of amber from Poland." Tourists are requested to buy amber only from these government shops. If commercial dealers purchase amber while in Poland from sources other than Cepelia, they are required to pay 80 percent duty when leaving the country. No raw amber may be exported. Therefore, most purchases are directly from Cepelia official government outlets with duties already added to the price. One American firm, the Amber Guild, has had a contract with Poland for many years for the export of raw materials and supplies Cepelia shops in the United States with finished amber products. Both tourists and dealers may make purchases from local artists in Poland if they have the finished pieces appraised and pay duties levied on the articles by the government officials upon leaving the country. Shopping from local artists often results in obtaining more creative articles than can be found in the usual production pieces. For example, a firm in Florida called "Amber Forever," imports ornate sterling silver jewelry cast by the lost wax method and set with amber. In some pieces, amber is combined with turquoise, amethyst, aquamarine or other semi-precious stones, and the Amber Forever collection contains unusual natural varieties of amber with rare color shades as well as the more typical transparent types.

Because of the shortage of raw amber, many factories in the amber centers of Gdańsk and Krolewiec now manufacture artificial amber articles and souvenirs. Synthetic resin is used to embed small natural amber chips in a product, called "polybern," which now is flooding the market in the

---

* Sterling silver is 0.925 pure silver, but much silver from European countries is only 0.869, unless stamped sterling silver.

Fig. 3–41. Cast silver and transparent amber pendant from Poland. *(Courtesy Amber Forever, Florida.)*

form of beads, brooches, vases and sculptures. Various jewelry articles and ornaments are also made of pressed amber (ambroid).

THE AMBER BELT OF DENMARK AND THE FRISIAN ISLANDS

Along the western coast of Jutland near Oksby, the present day "amber belt" of Denmark, after stormy nights, just as the early morning tide ebbs from the dunes, villagers still comb the beaches looking for amber lumps. Amber collectors of all ages, carrying lanterns to light their way, may be seen in the pre-dawn darkness searching the strand. On a "good" morning, that is, after a severe storm, there may be as many as a hundred collectors assembled from the nearby villages. As the generally small pebbles of amber are found, they are tucked safely away into leather pouches tied to the collectors' waists. Occasionally, a lucky searcher may discover a fist-size piece. But even after a storm, the total collection combined from several searchers in the Oksby region might only amount to as much as 22 pounds (10 kilograms). Most of these amber-collectors are amateurs, collecting amber for their own small-scale uses; however, a few native amber craftsmen make a living from collecting their own amber and cutting and selling amber jewelry. So little native amber is available that exports to the United States consist of about 80 percent of that which has previously been imported into Denmark from Poland, the Soviet Union and Germany.

Historically, the west coast of Denmark, as well as adjacent German

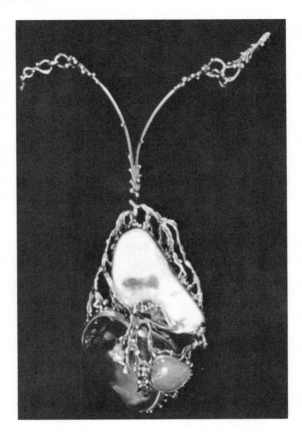

Fig. 3–42. Cast silver, white amber, yellow fatty amber and chrysoprase pendant made in Poland. *(Courtesy Amber Forever, Florida.)*

lands, comprise the region where the earliest Bronze Age amber trade route to the Mediterranean originated. The earliest mention of this region seems to be Pliny's statement: "the Gutones, a people in Germany, inhabit the shores of an estuary . . . one day's sail from this territory is the isle of Abalus upon the shores of which amber is thrown by the waves in the spring . . . the inhabitants use it for fuel and sell it to their neighbors, the Teutons." Some historians believe that Helgoland, the only island in the North Sea a day's sail from the German coast, is the classical "Isle of Abalus." Furthermore, the Eridanus River, referred to by classical writers, is thought to be the Elbe,* which empties into the North Sea in this region. The "Elektrides," or "amber islands" mentioned by the early Roman writers, were perhaps the Frisian Islands.[62]

Amber was much more abundant in this region in the past than at present. During the 1800's, for example, approximately 3000 pounds (1360 kilograms) of amber were collected per year along the North Sea coast of Denmark. The Islands of Romo, Sylt and Fanö produced amber, with some

* Some believe it to be the Po River.

lumps weighing several pounds each. During the period from 1822 to 1825, which included many stormy and therefore productive years, a Danish merchant reported collecting 686 pounds (311 kilograms) of amber along the shore near Ringkjöbing.[63]

Just across the narrow opening to the Baltic Sea, the southwest promontory of Scandinavia—including the stretch from Copenhagen to Malmo, Sweden—provides minor amounts of amber which can be picked up along the coast and in the waters of the Baltic after storms. Small nodules are most often found, but these are of good quality and are put to use by local artists.

The Zoological Museum of Copenhagen contains a large collection of insect-bearing amber specimens, many of which were collected from the shores of Denmark. This comprises one of the largest collections outside of Poland and Lithuania.

The most important Danish wholesale amber exporter, Einer Fehrn A/S,* buys much of his raw amber from the Soviet Union rather than depending on the local supply. Amber imported from Scandinavia includes excellent selections of natural amber beads, as well as articles made from reconstructed amber and jewelry using amber in typical Scandinavian styles and set in either gold or silver mountings. Major importers in the United States are Agneta Baltic Amber and Gifts from Scandia, both located in California.

WEST GERMANY AMBER LAPIDARIES

Beginning in the Middle Ages, skilled amber craftsmen established themselves in Germany away from the main source of raw amber, and this ancient industry still flourishes. Today, West German lapidaries manufacture fashionable amber beads and other goods which are characterized by fine quality and exquisitely fine finish. Many import-export firms deal exclusively in natural Baltic amber, and large quantities of amber are imported from the Soviet Union to provide raw material for the several firms scattered throughout the country and to the long-famous amber lapidaries located in Idar Oberstein, the well-known center of the German gem industry.

Although many amber necklaces are made from tumbled amber pieces, West German factories still produce hand-finished beads in round and olive shapes or faceted pendants and faceted beads. The latter, however, are becoming much more difficult to obtain than in the past because the skilled hand-work that is necessary in their production sharply increases costs. Amber specialists, such as Westfalica Amber Jewelry, Otto May and Gunther Herrling, mount amber in gold settings as well as in the typical silver mountings often used in Poland and the Soviet Union.

West German craftsmen continue the old traditions of carving amber, but on a lesser scale than in the past. Small display ornaments in the form of animals and flowers are carved in fine detail and require painstaking work

* Einer Fehrn A/S, Edisonvej 3, 1856 Copenhagen V.

Fig. 3-43. Carved bird of cloudy yellow natural amber from West Germany. *(Courtesy Kuhn's Baltic Amber, Florida.)*

on the part of the artist. Large lumps are carved in a manner that utilizes the piece to best advantage because large pieces of raw amber are now scarce.

The Natural History Museum in Berlin houses one of the largest single lumps of amber ever unearthed in East Prussia. This fine colored specimen weighs over 21 pounds (9.5 kilograms) and was found in 1860. In 1958, it was valued at over $4200, and it is much more valuable today.

Amber carvings and amber-veneered objects made during fifteenth, six-

Fig. 3-44. Carved transparent amber horse figurine from West Germany. *(Courtesy Kuhn's Baltic Amber, Florida.)*

Fig. 3–45. Large amber necklace (weighing 12 ounces). *(Courtesy Kuhn's Baltic Amber, Florida.)*

teenth and seventeenth centuries can be found in museums and churches throughout West Germany.

An American firm dealing exclusively in Baltic amber from West Germany is Kuhn's Baltic Amber. This firm possesses many unusual and exquisitely finished pieces, and exhibits two magnificent amber wedding necklaces, one weighing 12 ounces (340.2 grams) and another weighing 11 ounces (311.8 grams). The largest beads are as large as golf balls (see Fig. 3–45).

ENGLISH AMBER

Amber occurs sparingly at many localities along the east coast of England, but mainly along the North Sea shores of Norfolk and Suffolk. Seaside resorts in Aldeburgh, Cromer, Felixstowe, Freat Yarmouth, Lowestaft and Sathwold specialize in amber products since they are located along the area where amber is occasionally found. Local shops carry fine amber jewelry imported from West Germany, Denmark, Poland and the Soviet Union, as well as hand-crafted pieces of local origin.[64]

English amber is usually golden or cloudy yellow and generally occurs in small lumps, although large pieces of up to 2 to 3 ounces have been reported. English amber is classed as succinite, but it commonly contains shells and otherwise reveals different types of flora and fauna from those of the amber in Samland. Its source is not known, but presumably it has washed from a Tertiary bed now submerged under the North Sea.

Amber artifacts found in prehistoric graves in England evoked disagreement among scholars as to whether the amber originated in England before it was worked or was imported from the continent. Since English and Baltic amber both contain succinic acid, one can only speculate whether Stone Age beads and ornaments came from English or Baltic shores. Most Bronze Age sites in England are in Wessex, where the Wessex Culture left evidences of its existence in the remains of their now-famous monument, Stonehenge. The barrows or burial mounds of this period often contain gold, ivory and amber. Being sun-worshippers, these ancient people believed amber was related to the sun, and for this reason, amber pendants set in a gold circle were worn as solar amulets.[65]

One of the most unusual amber artifacts found in a Bronze Age tomb (1500 B.C. to 1150 B.C.) in Dorchester is an amber cup of a reddish golden color. The cup measures 2.5 inches (62 millimeters) high and 3.5 inches (87 millimeters) wide. It was formed from one piece of amber. The surface is smooth and seems to have been turned on a lathe, but its solid handle was carved from the original (and remarkably large) piece of amber. The cup was discovered in 1857 during construction of a railway station, when workmen uncovered an old grave and found a crude coffin, hollowed out of a tree trunk. In the coffin was the amber cup, along with an axehead of obsidian and another stone amulet. What better way to immortality than an axe for protection on one's journey to the spirit world, and an amber cup to drink ambrosia with the gods![66]

In a later English period, Anglo Saxon craftsmen discovered that amber could be easily carved and polished. It had abundant uses and was believed to possess the power to protect its wearers against witchcraft and evil spirits. It was unnecessary to wear an entire necklace of amber, since a single piece sufficed.[67]

While studying artifacts from sixth century graves in England, archaeologist T.C. Lethbridge observed that amber was more common in graves dated to the second half of that century. He speculated that the increase was a result of improved trade, possibly with the inhabitants of the North Sea coast. Abington Cemetery, which lies along one of the major English trade routes, yielded graves that dated later than the sixth century and also contained amber in more abundance than in earlier graves, possibly indicating not only an increase in availability, but an increasing esteem based on amber's magical properties.[68]

As early as 1450, English literature, in the *Book of Curtasy,* indicated the importance of amber beads for ladies' wear. During the sixteenth century, the celebrated English poet and writer, Francis Bacon (1561 to 1626) described amber inclusions in his *Histories of Life and Death:*

The Spider, Flye and Ant, being tender dissipable substances falling into amber are therein buryed, finding therein both a Death, and Tombe, preserving them better from Corruption than Regal Monument.

Sir Thomas Moore (1779 to 1852), the poet of Ireland, was also intrigued by amber, and, referring to Sophocles' belief that amber was produced by the mourning of mythical birds, wrote:

Around thee shall glisten the loveliest amber
That ever the sorrowing sea-bird has wept;
With many a shell, in whose hallow-wreathered chamber
We, Peris of Ocean, by moonlight have slept.
        The Lament of the Peri for Hinda, in *The Fire Worshippers,* verse 8.

Other English poets also were fascinated by amber. References to the gemstone were included in the poetry of Christopher Marlow, Sir Walter Raleigh, Pope and Tennyson. Tennyson's *Lover's Tale* mentions lovingly:

The loud stream,
Forth issuing from his portals in the crag
(A visible link unto the home of my heart)
Ran amber toward the west.

In fact, John Milton was so impressed with amber when writing *Paradise Lost, Book VI,* he portrayed the "Chariot of the Paternal Diety" as "inlayed with pure amber."

During the period shortly after World War I, amber enjoyed a renewed popularity in England, and beaded necklaces were very much the fashion. Being lightweight and warm to the touch, long necklaces of faceted amber beads were worn with great comfort. The more the beads were worn, the more beautiful they became as they mellowed with age.

A long-established amber shop of London, Sac Freres, is to be found on Old Bond Street. Their specialty has been amber for many decades, and this shop is well worth a visit for connoisseurs traveling in England. Elsewhere in London is the Albert and Victoria Museum, which contains several carved altar pieces and jewelry caskets from the sixteenth and seventeenth centuries. Because amber has long been a favorite in English jewelry, antique shops often have amber articles for sale.

The largest known piece of amber found to date is housed in the Natural History Museum in London. The piece weighs 33 pounds, 10 ounces and is thought to be Burmese amber. It was acquired by John C. Bowing in 1860 for £300 in Canton, China and was placed in the museum in 1940.

REFERENCES

1. Haddow, J. G., *Amber, All About It*. Liverpool, England: Cope's Smoke Room Booklets, No. 7, 1891, p. 26.
2. Wiliamson, G. C., *The Book of Amber*. London: Ernest Benn, 1932, pp. 98–99.
3. Ley, Willy, *Dragons in Amber*. New York: The Viking Press, 1951, p. 13.
4. Spekke, Arnold, *The Ancient Amber Routes and the Geographical Discovery of the Eastern Baltic*. Stockholm: M. Goppers, 1957, p. 10.
5. Williamson, *The Book of Amber*, p. 99; Spekke, *The Ancient Amber Routes*, p. 9.
6. Williamson, *The Book of Amber*, p. 103.
7. Ley, *Dragons in Amber*, p. 15.
8. Haddow, *Amber, All About It*, p. 38.
9. Ley, *Dragons in Amber*, p. 16.
10. Williamson, *The Book of Amber*, p. 104.
11. Pelka, Otto, *Bernstein*. Berlin: Richard Carl Schmidt, 1920.
12. Ley, *Dragons in Amber*, p. 18.
13. Gudynas, P. and Pinkus, S., *The Palanga Museum of Amber*. Vilnius: Mintis Books, 1967, p. 94.
14. Williamson, *The Book of Amber*, p. 108.
15. Hunger, Rosa, *Magic of Amber*. London: N.A.G. Press, 1977, p. 62.
16. Suprichev, Vladimir, [Amber — Talisman, Medicine, Ornament.], *Nauki i Tekhnika*, December 1978, pp. 20–22.
17. *Ibid.*
18. Hartmann, Philip, *Succini prussici physica & civilis historia*. Frankfurt: 1676.
19. Bauer, Max. *Precious Stones* II: 540. London: Charles Griffin & Co., 1904; reprinted by Dover, 1968.
20. Haddow, *Amber, All About It*, p. 34.
21. Bauer, *Precious Stones* II: 541.
22. Ley, *Dragons in Amber*, pp. 23–24.
23. Anonymous, "Diving for Amber," *St. Paul's Magazine,* ca. 1890.
24. Ley, *Dragons in Amber*, p. 26.
25. Haddow, *Amber, All About It*, p. 35.
26. Ley, *Drag ns in Amber*, p. 17.
27. Williamson, *The Book of Amber*, pp. 127–128.
28. Ley, *Dragons in Amber*, p. 33.
29. Bauer, *Precious Stones* II: 543.
30. *Ibid.,* II: 544.
31. Zaddach, G., *Das Tertiärgebirge Samlands*. Königsberg: Schriften der Physikalishe-Ökonomische Gesellschaft Jg. 8, 1867, pp. 85–197.
32. Ley, *Dragons in Amber*, p. 34.
33. Haddow, *Amber, All About It*, p. 37.
34. Ley, *Dragons in Amber*, p. 35.
35. Jakubowski, M., "Amber Fishers and Mines," *The Jewelers' Circular-Weekly,* December 23, 1914, p. 14.
36. Ley, *Dragons in Amber*, p. 35.
37. Bauer, *Precious Stones* II: 550.
38. Bellmann, E., "Amber," *The Mining Journal* 101: 129, 1913.
39. Bellmann, E., "Recovery and Treatment of Amber at Palmnicken (East Prussia)," *The Mining Journal* 102: 722, July 26, 1913.
40. Jakubowski, M., "Amber Fishers and Mines," *The Jewelers' Circular-Weekly,* December 23, 1914, p. 14.
41. Bellmann, E., "Recovery and Treatment of Amber at Palmnicken (East Prussia)," *The Mining Journal* 102: 722, July 26, 1913.

42. Petar, Alice V., "Amber," *Information Circular 6789*. United States Bureau of Mines, June 1934, p. 3.
43. Weinstein, Michael, *The World of Jewel Stones*. New York: Sheridan House, 1958, p. 228.
44. Petar, "Amber," p. 6.
45. Kraus, Edward H., *Gems and Gem Materials*. New York: McGraw-Hill, 1939, pp. 239-240.
46. Suprichev, Vladimir, [Amber—Talisman, Medicine, Ornament], *Nauki i Tekhnika,* December 1978, p. 21.
47. *Ibid.*
48. Petar, "Amber," p. 8.
49. Baer, John W., "Floating 'Gold' from the Baltic," *The Jewelers' Circular-Keystone,* October 1937, pp. 82-86.
50. Prockat, Friedrich, "Amber Mining in Germany," *Engineering and Minerology Journal* **129:** 305-307 March 24, 1932.
51. Petar, "Amber," p. 7.
52. Kemp, Edwin C., *American Consular Commerce Reports*. Danzig: American Consulate, April 27, 1925, p. 212.
53. Petar, "Amber," p. 8.
54. Ley, *Dragons in Amber,* p. 35.
55. Weinstein, *The World of Jewel Stones,* p. 225.
56. Ley, *Dragons in Amber,* p. 36.
57. "Amber," *Encyclopedia Lituanica*. Boston, 1970, pp. 84-87.
58. Almazjuvelirexport, *Bernstein,* Moscow, 1978, p. 1.
59. Suprichev, Vladimir, [Amber—Talisman, Medicine, Ornament], *Nauki i Tekhnika,* December 1978, p. 22.
60. *Ibid.*
61. Szejnert, Malgorzata, *Traffic on the Amber Route,* Poland, 1977 (Source unknown).
62. Ley, *Dragons in Amber,* p. 11.
63. *Ibid.,* p. 12.
64. Weinstein, *The World of Jewel Stones,* p. 226.
65. Hunger, *Magic of Amber,* p. 22.
66. Williamson, *The Book of Amber,* pp. 82-84.
67. Bradford, Ernle, *Four Centuries of European Jewellery*. Middlesex, England: Spring Books, 1953, p. 209.
68. Hunger, *Magic of Amber,* p. 28.

ADDITIONAL REFERENCES

Block, J. A., *The Story of Jewelry*. New York: William Morrow, 1974.
d'Aulaire, Emily and Ola, "For Ever — Amber," *Reader's Digest* (reprint from Scandinavian Edition), December 1974.
d'Aulaire, Emily and Ola, "Amber: Gold of the North," *Scanorama Magazine,* January 1978, pp. 61-64.
Farrington, Oliver C., "Amber," *Department of Geology Leaflet No. 3*. Chicago: Field Museum of Natural History, 1923.

4

The Curious Lore of Amber

Oh, listen in the evenings,
When the sea is restless
And sprays the shore with amber
The depths unseen palm . . .
Maironis

ANCIENT MYTHS

For many centuries, cultures acquainted with amber wondered what this beautiful gift of nature was and how it originated. True to the customs of the times, many myths and legends relating to amber resulted from attempts to explain the origin of this mysterious substance. Most of the folklore was simply carried from generation to generation by word of mouth, as older members of the culture shared their wisdom with youth. Some legends found their way into the literature of antiquity and are preserved for us today in their poetic versions.

It was the Roman poet Ovid who recorded the prevalent Greek myth, the *Tears of Heliades,* attributing divine origin to amber. The Greek legend recounts the adventures of Phaëthon, who grew to young manhood without knowing that one of his parents was immortal. Prevailing upon his mother, Clymene, to inform him who his father was, Phaëthon was astonished to discover that he was the child of the sun god, Helios. Doubting his mother's word, he sought out the Sun God to seek proof of his parentage.

Phaëthon journeyed to India, where Helios' palace stood splendidly, glimmering with precious stones and gleaming gold. Upon entering the hall, the boy was blinded by light from the Sun God and found it impossible to see. But upon seeing Phaëthon, Helios dimmed his radiance and commanded Phaëthon to approach.

Phaëthon demanded: "Helios, if you truly are my father, what proof will you give so I may be known as your son?"

"You deserve not to be disowned, my son. Whatever you ask I will grant you," promised Helios.

Thereupon the boy asked to be permitted for one day to drive the flaming chariot of the sun across the arch of the heavens.

Knowing the great dangers involved, Helios offered Phaëthon other gifts, attempting to persuade the would-be charioteer of the skill necessary and the dangers to be encountered. But possessing the willfulness of youth, Phaëthon could not be deterred. With great reluctance, Helios gave his consent.

On the appointed day, the daughters of Helios, the Heliades, helped Phaëthon yoke his father's steeds to the chariot. As dawn opened the doors of the east with the stars and moon retiring, Phaëthon leaped into the golden chariot. Delightedly grasping the reins of the fiery horses, he sped off to the west. Soon the palace was far behind him as the impatient horses sprang swiftly forward, outrunning the morning breezes.

Feeling a lighter load than usual, the horses soon realized that inexperienced hands held the reins, and they ran wild, straying far from the traveled path. Seized with fear, poor Phaëthon forgot his horses' names and

Fig. 4-1. Natural amber mosaic.

knew not how to guide them. He dropped the reins in terror on the height and became overwhelmed with dizziness. The uncontrolled steeds dashed on without restraint, rushing hither and thither, wherever they chose. As they approached too near the Earth, clouds began to smoke, the harvest blazed, fields were parched and the rivers dried up. Even the sea shrank because of the awful heat.

Fearing the world would be reduced to ashes, the Goddess of Earth begged Jupiter to take pity and give her relief. In an effort to save the Earth from being consumed, Jupiter launched a thunderbolt at Phaëthon, causing him to fall headlong and flaming into the Eridanus River. Finding him dead, the river nymphs reared a tomb to Phaëthon's memory along the shore.

Lamenting Phaëthon's untimely end, his sisters, the Heliades, accompanied by their mother, a daughter of Oceanus, wept bitterly. As punishment for encouraging Phaëthon's reckless ride and for assisting him in yoking the steeds to the chariot, the Heliades became rooted to the spot and changed into poplar trees, from which tears continually fell. These tears became hardened by the sun and turned into amber.[1,2]

Fig. 4-2. Phaëton driving the Chariot of the Sun (From title page of Haddow, *Amber, All About It,* 1891.)

Fig. 4-3. The orgin of amber. The Heliades weeping over the untimely death of their brother. (From Haddow, *Amber, All About It,* 1891.)

This early Greek myth poetically illustrates that the relationship between trees and amber was observed even in antiquity. In an attempt to explain the naturally occurring annual phenomena of the scorching hot summer sun, when the Earth seems about to be burned, the ancients blamed the unskilled driver of the sun chariot. As fall approaches, the hot spell ends; thunderbolts cross the sky in a siege of lightning and rain, indicating that Mother Earth's appeal had been heard. Phaëthon's sisters' taking root as poplars weeping tears of amber shows that though some ancients believed amber to be a product of trees from a river area, they confused the source, relating amber to the black poplar tree rather than to a coniferous tree.[3]

Other ancient philosophers and poets of the era were not as observant as the creator of this myth and took different views regarding the origin of amber. Nicias of Greece, for example, regarded amber as a liquid produced by rays of the sun striking the surface of the soil with tremendous force, producing an "unctious sweat" that was carried by waves to the sea. According to Nicias, it then solidified in the water and was thrown back on the shores of "Germania."[4] Another fanciful tale explained amber to be honey melted by the sun and dropped into the sea from the mountains of Ajan, whereupon it became congealed by the water.[5]

The Greeks were not the only people who identified amber with burned honey. Far to the east in ancient China, hundreds of years before the beginning of the Cristian Era, a Chinese scholar, who may have studied the translucence of amber in sunlight and perhaps had even obtained a piece with a bee preserved within, recorded the following myth: "Somewhere are cliffs, the cliffs of Ning Chou, in which dwell thousands of bees. When the Cliffs crumble, the bees come out. People burn them and make them into amber."[6]

Other ancient European myths involved amber being gathered in the Gardens of the Hesperides, where golden apples of immortality were guarded. This belief was influential enough so that amber itself was bestowed with powers of immortality.

An old herbal, the *Hortus Sanitatus,* published just before Columbus discovered America, contains a quaint old woodcut of "the tree that exudes amber"(see Fig. 4-4). The tree is depicted with enormous drops of sap generously flowing from the trunk of a gnarled tree of some fabulous species.[7]

Sophocles, possibly having received one of the few pieces of amber with feather inclusions, believed that amber was produced in the countries beyond India from the tears shed for the hero Meleager by his sisters. According to the legend, Meleager's sisters were transformed into birds (called "meleagrides") who left Greece once a year, flying beyond India and weeping tears which turned into amber.

Early classical cultures were not the only ones to produce philosophers who attempted to provide an explanation for the origin of amber. The Aisti, or early Lithuanians, having immediate experience with the amber nodules that were thrown onto their sandy shores by the Baltic, also created magical

Fig. 4-4. The tree that exudes amber. (From *Hortus Sanitatus,* 1491.)

legends about amber's origin. Amber-related myths, born in Lithuania and inspired by the restless Baltic, are in their own way as poetic as those originating in the Classical Mediterranean cultures. In one Lithuanian legend, amber is portrayed as the fruit of a passionate and tragic love. According to legend, during ancient times, Perkunas, God of Thunder, was the father God. The fairest of all goddesses was Juraté, a mermaid Goddess of the Sea, who lived in an amber palace at the bottom of the Baltic. Kastytis, a courageous fisherman living along the Baltic coast near the mouth of the Sventoji River, would cast his nets to catch fish from Juraté's kingdom. Displeased by this intrusion, Juraté sent her mermaids to warn Kastytis to leave her fish alone and disturb the sea no more. Paying no heed to her warnings and impervious to the charms of her mermaids, Kastytis continued to cast his nets and bring in fish. Watching the fisherman haul his catch into his boat, Juraté saw how handsome Kastytis was and admired his great courage. Possessing human failings, Juraté fell in love with the mere mortal, Kastytis, and, in spite of great differences between them, Juraté took the fisherman to her amber palace.

Perkunas, knowing Juraté was destined to be the consort of Patrimpas, the God of water, became greatly angered upon discovering the immortal goddess in love with a mortal. In his fury, the God of Thunder sent a shaft of lightning from the skies, striking Juraté's palace, demolishing it into thousands of fragments and killing her lover, Kastytis. Juraté, crying tears of amber for Kastytis and their tragic love, was punished by being chained to the ruins of her castle. Thus, it is said that when storms churn the cruel Baltic, Juraté is being tossed to and fro by the waves—and even to this day, while storms are raging at sea, the sound of Juraté wailing in the depths may still be heard as she mourns for Kastytis, a son of the earth. As her weeping becomes emotional, the peaceful depths of the sea grow restless and stormy and lumps of amber from her demolished palace are spewed up from the sea bottom. Becoming entangled in seaweed, the amber lumps are thrown out onto the Baltic shores.

To the Lithuanians, the small, tear-shaped pieces of amber are the tears of Juraté, as clear and pure as her tragic love. The legend lives today through a variety of beautiful sculptures, mosaics in amber and incrustations of amber designed by Lithuanian artists.

Fig. 4-5. Amber and bronze sculpture depicting the legend of Juraté and Kastytis. (M. Kizis' collection.)

Though passed along by Balts for centuries, the Juraté and Kastytis tale was first recorded by Lindvikas Adomas Jucevicius, a historian of Lithuanian literature, in 1842. Many artistic works of poets, artists and composers in Lithuania have been inspired by the legend. Upon entering the Palanga Amber Museum, the visitors are transported into the immortal poetic world of Lithuanian folklore by a symphony based on the legend of Juraté and Kastytis. Music, poetry and visual art in the form of a bronze sculpture and a colorful stained glass picture introduce the visitor step by step to the origin of amber—the "Gold of the North."[8]

The Norsemen, another ancient race familiar with amber as a product of nature, relate their own version of the divine origin of amber, in the myth of Freya and Odur.[9] Freya, the beautiful, blue-eyed, blond goddess of love, beauty and fertility, had a great fondness for jewels. She was wed to handsome Odur, the sunshine, whom she loved dearly, and they dwelt happily in the land of Asgard in Freya's palace, Folkvanger, with their two lovely daughters. Little did Freya know that her love for jewels was soon to bring her great sadness.

One day, as Freya walked along the border of her kingdom where the Black Dwarfs dwelt, she spied them making the most wonderful necklace, glistening as bright as the sun. It was called Brisingamen, or the Brising necklace. Freya caught her breath on seeing the beauty of the golden necklace.

"Oh, please sell me this necklace for a treasure of silver," she begged. "I cannot live without it, for I have never seen one as beautiful."

The dwarfs replied: "All the silver in the world cannot buy the Brisingamen from us."

Believing her life could not be endured without the beautiful Brising necklace, Freya asked: "Is there any treasure in the world for which you would sell me the necklace?"

"Yes, you must buy it from each of us," answered the dwarfs, "for the treasure of your love. If you are wed to each of us for a day and a night, Brisingamen shall be yours."

Awed at the sparkle of the world's loveliest adornment, Freya was overcome with madness. She forgot Odur, her beloved husband. She forgot her two fair daughters. Indeed, she even forgot she was Queen of the Aesir! She agreed to the pact. No one in Aesir knew about the dwarf weddings except the mischief-maker, Loki, who always seemed to know where evil was brewing.

After four days and nights of these unholy unions, Freya returned to Asgard to dwell again in her palace. Ashamed of what she had done, she hid the shining golden necklace as she went to her chambers. But Loki sought out Odur to inform him of the happenings in the country of the dwarfs, whereupon Odur demanded Loki produce proof of his tale. To do so, Loki set out to steal the Brising necklace. Turning himself into a flea, he flew into Freya's chambers, and, finding her asleep, he bit her on the cheek, causing her to turn so that he was able to unfasten the necklace. Carefully slipping the Brisingamen away from Freya, Loki went straight to Odur to show him the necklace.

Odur, upon seeing the necklace, tossed it aside and wandered out of Asgard into the world, far into distant lands.

In the morning, Freya woke to discover Brisingamen gone. Intending to tell Odur, she sent for him—to find that he was gone as well. Weeping bitterly, she went to Valhalla to confess to Odin, the father god, whose palace was near the amber valley paradise of Glaesisvellir.* At the gates of Valhalla was an amber grove called Glaeser, with trees that dripped glistening amber.

Odin forgave Freya for the evil she had done, but taking the necklace from Loki, he decreed she must wear Brisingamen forever to remind her of the past. Fastening the Brising necklace around her neck, Freya went forth into the world in search for Odur. As she wandered, she continued weeping. The teardrops falling on the land turned to gold in the rocks, while those falling into the sea turned to drops of amber.

In Baltic countries, the belief is that an amber necklace chokes the wearer who tells an untruth, reminding one not to do evil. It is perhaps from this source that amber, in the language of gems, became known to mean "disdain."[10]

The folklore of the Lithuanians contains another imaginative tale, *Amberita,* written by J. Narune,[11] which explains why amber is tossed on their shores as a gift from the generous sea.

The legend says that living near the sea was a family: an old fisherman, his wife, and their lovely daughter, Amberella. Amberella had long golden hair and deep sea-blue eyes, and she would sing like a nightingale as she played on the beach. One day, while swimming, she found a large piece of beautiful amber, but as she grasped the lump in her hands, she felt a strange power pulling her. She tried to swim away, but it was impossible. She was tossed by a wave and pulled into a whirlpool, and a sweet drowsiness came over her as she sank deeper into the sea. As if in a dream, Amberella felt someone grasp her and heard a hard, steely voice say: "Now you are mine!"

The poor old fisherman and his wife were grief-stricken at the loss of their only daughter. Amberella's mother searched the seashore, calling her little girl's name and crying. A goldfish heard the sorrowful cries, and taking pity, she pleaded: "Weep no more. Amberella lives at the bottom of the sea in a beautiful amber palace. The Prince of the Seas has captured her for his wife and he loves her."

The goldfish offered to bring pearls or lumps of amber to cheer the longing couple, but the old woman said sadly: "We want our darling Amberella back, not pearls or amber. Go tell her how our hearts grieve and how lonely we are without her."

---

* The words, Glaeser and Glaesisvellir, were derived from the old German word for amber, *gles* (glaes), which later, when glass was introduced into Northern Europe, became the word for glass (see nomenclature).

Fig. 4–6. Amber mosaic of Amberella. (V. Tomaslunas' collection.)

The goldfish felt sorry for the lonely old woman, and as she disappeared into the depths, she promised to try.

Descending deeper and deeper into the water, the goldfish searched for the Prince of the Seas' amber palace. Escaping many perils along her journey, she at last arrived at the palace. What splendor sparkled before her eyes! Towers of gleaming red and golden yellow amber overwhelmed the goldfish. The glistening amber gate was guarded by two large lions carved of yellow amber, while between the lions an ugly octopus checked each visitor before entry. Though frightened and tired, the goldfish told the octopus she had an important message for the Prince of the Seas.

Inside the palace, the goldfish gazed with awe at the wonderful sight of branches of amber blossoms glowing in the deep waters and casting a strange, mellow light. The walls sparkled like diamonds with tiny chips of amber. Hanging from the ceiling, the amber teardrops were like droplets of honey. White foamy amber formed wreaths around the transparent amber columns which supported the palace. The glistening hallway was lined with a row of goddesses carved of precious green amber. Never having seen anything so beautiful, the little goldfish was filled with delight.

At the end of the hall, sitting on an amber throne as bright as the sun, the Prince of the Seas sat, with Amberella beside him. When the goldfish had

Fig. 4–7. Carved cloudy yellow amber lion from West Germany. *(Courtesy Kuhn's Baltic Amber, Florida.)*

delivered her message, Amberella begged the prince to let her go to visit her dear mother.

The prince gave out a terrifying roar. The sea echoed his wrath. He summoned the Icy Wind to blow and toss the sea about. Huge waves bellowed and spouted foam. Suddenly, from the Dark Waters, white foaming horses appeared with heads held high, manes flying in the air. Upon the finest of all the steeds sat the Prince of the Seas, holding Amberella in his arms as they sprang out of the water.

Atop Amberella's flowing hair glistened an amber crown. A lovely amber necklace adorned her dress, and in her hands she held large chunks of sparkling amber. On the shore, the old fisherman and his wife gazed in amazement at their beautiful daughter glowing with amber.

"Come back, Amberella. Come back!" begged her mother.

Amberella was held securely by the prince, but she hurled large lumps of amber onto the shore.

The Prince of the Seas made a sign, and suddenly the storm subsided. Amberella, the white horses and the prince vanished into the sea.

The old fisherman and his wife sadly realized that Amberella would never come back to their cottage again. Bowing their heads with grief, they saw the amber lumps thrown to them by their daughter.

Now, whenever the Prince of the Seas becomes angered, the sea begins to storm. Amberella tosses amber from the sea to her parents to show them how much she misses and loves them. Longing for her own cottage again, she brings amber of all sizes and colors from her palace under the sea. Many times after a storm, the entire length of the strand is covered with amber.

POWERS AND SYMBOLISMS OF AMBER

O! Mickle is the powerful grace that lies
In herbs, plants, stones, and their true qualities.[12]

Amber not only was thought to protect the living, but it was believed to speed the dead on their journey into the shadows. In ancient Scandinavia, for example, an amber axe (see Fig. 4–8) was placed in tombs along with other treasured possessions to protect the soul during its journey and to confer immortality. The word designating the fabled drink of immortality, *ambrosia,* is cognate to amber, as is the Greek word *ambrotos,* meaning immortal.

The association of amber with immortality continued in the older folklore of Eastern Europe, the British Isles and Scandinavia, with amber mountains representing the land where the dead dwelt at the end of the world. These mountains, as well as amber islands, appear to be the forerunners of the glass mountains and islands so commonly found in fairy tales. Because amber in early times was called "gles" in northern regions of Europe, it became confused with glass in stories when this same word began to be used

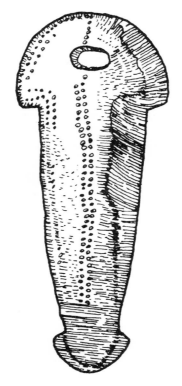

Fig. 4–8. Neolithic amber axe, 3 inches long, found in the Vistula River in Poland. (From Strums, *Die Neolithische Plastik,* as in Spekke, *Ancient Amber Routes,* 1956.)

to designate the latter. Therefore, as time passed, amber mountains and islands were forgotten, and instead they became glass mountains and islands.

As tales were passed from generation to generation, the climbing of the amber mountain by the characters in the story often developed into a test for the hero to pass to win his princess. The land of the dead of early tales later became transformed into a magical place of protection or safety where the princess was safeguarded for immortality or until the hero was able to win her. Since in the fairy tales, witches were often characterized as living on such mountains, amber amulets were used as a protection against evil witchcraft. This practice spread as far south in Europe as Italy.[13]

Especially prized for its imagined powers of protection against witchcraft was an amber pendant enshrining a small insect. Thus, such a talisman brought an enormous price.[14] This belief was so widespread that during Pliny's time, for instance, even young children wore amber around their necks to protect them from the evil influence. Ancient legends relating accounts of witchcraft also attribute to amber "the power to warn its masters, the favored of the witches, of their danger and when handled by its master to turn pale, to gleam with rippling light, and to emit a perfume."[15]

Spindle whorls for spinning thread were often made of amber as a protection against the evil spirits which were "known" to put hexes on thread as it was spun, causing it to snarl. Amber was also used for sewing and crocheting implements as further protection against this form of witchcraft.

The most powerful protection against the evil eye was a phallus made of amber. This was regarded as protection against any and every attack of evil

Fig. 4–9. Amber heart engraved with the design of a bird. Engraved gems were attributed more magical power than the stone alone. (Modern piece by scrimshaw artist Bryce Barker.)

spirits. Amber amulets were also made in the form of lions, dogs, rabbits, frogs and fish in Eastern Asia and were believed to add to the virility of men and the fecundity of women.[16] Cut into various other magical forms, amber was used widely in Baltic, Scandinavian and Mediterranean cultures to afford similar protection against the evil eye, witchcraft and sorcery. In the Baltic regions of Germany, Denmark and Sweden, where gathering amber was not highly restricted, peasants carved amber hearts during long cold winter evenings while sitting by the fire. These were given to their sweethearts as a sign of their love and to keep them free from harm. What better way to prove one's love? Scandinavian peasants originally used amber buttons as lover's gifts, talismans against evil that might befall the loved ones. These too were fashioned by home artisans and can be found in a variety of unusual and symbolic shapes. Custom decreed that amber be worn by brides to insure happiness and long life.

Illustrating amber's symbolism of courage, "The Torch of Happiness," a Baltic folk legend, is a tale describing the enduring courage of a youth who was not deterred by seemingly impossible conditions. The tale is often exquisitely depicted in amber mosaics from Lithuania, in which a young boy is protrayed triumphantly holding a lustrous amber torch above his head as he stands on top of an amber mountain, entirely represented by natural variations in the shape and color of the amber itself. The legend tells of a group of very poor people living at the foot of a steep hill, at the top of which was the flickering "Torch of Happiness." Villagers could see the torch's shining rays, but it was out of their reach. If they could only reach the torch, happiness would be theirs.

One by one, the brave villagers attempted to climb the almost vertical sides of the hill. As hard as they tried the slope overcame them and they fell to the ground. When the "would-be" rescuers failed, they were turned to stone, each adding to the rugged climb for the next person. Finally, a young boy of great courage, knowing full well if he did not succeed he too would be turned to stone, tried to scale the hillside. Carefully placing each footstep as he struggled over the boulders, he continued his perilous climb, and, because of his splendid effort, he was soon within view of a glowing flame. In disbelief that he had actually succeeded in climbing to the top, he reached out for the torch. Grasping it in his hands, the boy watched with wonder as the stones of the unsuccessful climbers suddenly turned back into live persons. Surely, he had reached the "Torch of Happiness." Today, the amber "stones" found in the countryside serves as a reminder of the courage of this persistant youth. (See Color Plate 34.)

Far to the East, amber was also used as a symbol of courage. The ancient Chinese believed it contained the soul of the tiger, while in the Buddhist paradise, the purest souls are those with bright amber-yellow faces. Further merit may grow in the shape of diamonds, flowers and amber. Thus, amber was also given a religious significance.[17]

The ancient amber-gatherers viewed amber as sacred to the mother goddesses because life substance was concentrated within. Tacitus, the early

Fig. 4-10. Neolithic amber boar figurine found near Gdańsk, Poland (Danzig, East Prussia). (From Strums, *Die Neolithische Plastik,* as in Spekke, *Ancient Amber Routes,* 1956.)

historian of the Germanic tribes, described inhabitants on the Baltic southern shore as worshippers of "the mother of the gods." He informs us that "the figure of a wild boar is the symbol of their [the Balts'] superstition, and he who has that emblem about him thinks himself secure even in the thickest ranks of the enemy, without any need of arms or other mode of defense."

The amber boar, symbol of the amber goddess, was also the sacred animal of Celtic tribes of ancient Britain. The symbol of the boar, found on early British armor, became known in succeeding centuries as the "lucky pig."[18]

The Celtic sun father, Ambres, derives his name from amber. The

Fig. 4-11. Amber egg with engraved symbol of the crowing rooster, symbol of sun worship. (Author's collection; Bryce Barker artist.)

gemstone was used in sun-worship during Neolithic times and was often carved as an emblem of the sun. It has been found among Cornish megaliths or stone monuments, as the central pillar representing the sun as Lord of Time.[19] Another symbol related to sun-worship was the crowing rooster, who signaled the arrival of the sun each morning. The amber egg engraved with a crowing rooster appropriately represented the many virtues attributed to amber: new birth, new dawn, a new life substance given by the sun. (See Fig. 4-11.)

In the Old Testament, amber was the gem of the tribe of Benjamin. Amber was burned as an incense and it is believed its pungent perfume was one of four types Moses commanded for Tabernacle use.[20] The early Christian tradition included amber as a sign of the presence of God. It was worn in battle to protect the warriors and defend them against the enemy.

It is said the former Shah of Persia possessed a cube of amber which fell from heaven in Mahomet's time. The amber possessed the power to make the Shah invulnerable.[21]

TALISMANIC, AMULETIC AND FOLK MEDICINE

An amulet can be a powerful healer if the owner believes with sufficient faith in what it can do; for thought is sometimes a miracle-worker when the mind accepts the possibility of healing. And so faith, centered though it may be in an inanimate object, can flower to heal in a most wonderful way.

Yet I think there is something more, a blending of the physical and the mystic, which makes amber a potential agent for healing as it does all precious and semi-precious stones.[22]

Doris M. Hodges

Since time immemorial, amber has been worn as an amulet in the Baltic as well as the Mediterranean cultures. The people of these cultures believed amber was especially beneficial for protecting infants from the croup when worn as a necklace in the form of an amulet. As a charm, amber was believed to ward off fits, dysentery and nervous afflictions. As early as Pliny's time, farmers' wives in areas where the "water had properties that would harm the throat," or was deficient in iodine, wore amber beads around their necks as a remedy against swellings (goiter).

Surprisingly, the lore attributing guardian values to amber has lasted into the twentieth century. During the late 1960's, for example, an authority announced the virtues of wearing amber beads around the throat to protect it from diseases. It was believed that the strong electrical properties of amber resulted in an electrical band forming around the throat and bringing into existence a protective power.[23] Even today, in many of the peasant areas of the Baltic, amber earrings and necklaces are worn when one has a headache or a throat ailment.

Amber was thought to possess many other unusual curative powers. During Pliny's era, amber was powdered and mixed with honey and oil-of-roses for curing ear troubles; when mixed with honey from Attica, it was a cure for dimming eyesight. Since Hippocrates' time, or about 400 B.C., amber powder and oil-of-amber have been used in medicines. The powder was taken internally as a remedy for diseases of the stomach, whereas the oil, with properties resembling turpentine, was used in medicines prescribed for internal administrations for asthma and whooping cough. However, amber oil was more frequently used as a linament to be rubbed on the chest. Simply holding a ball of amber in the hands not only kept one cool during the hottest days of summer, but would reduce the temperature of a person suffering from fever.

During the Middle Ages, amber was used to ease stomach pains and goiters and was believed to act effectively against certain poisons. Even jaundice was believed to be cured by wearing amber. It was thought that the unhealthy color of the skin—and with it, the sickness—was extracted magically by the powerful yellow of the stone.

In 1502, Camillus Leonardus expounded the virtues of amber as a medicine in his *Speculum Lapidum*. He stated: "Amber naturally restrains the flux of the belly; is an efficacious remedy for all disorders of the throat. It is good against poison. If laid on the breast of a wife when she is asleep, it makes her confess all her evil deeds. It fastens teeth that are loosened, and by smoke of it poisonous insects are driven away."[24] During the rule of the Teutonic Knights, white amber was so highly valued for it medicinal uses that the rights for white amber were never sold when the amber monopolies were released. The Order had originated in 1190 as a charitable order in association with the German Hospital of St. Mary. The military character of the Order was assumed in 1198, and the medicinal uses of amber were held with great regard by the Knights.

In the mid-1500's, Duke Albert's court physicians published a scientific treatise relating the specific uses of white amber. Having been used as a talisman for centuries, this amber gained scientific credibility during this mystical era.

Many prescriptions were written which included amber along with other gemstones. An old meteria medica lists 200 medicinal stones. A dose suggested for heart disease contained white amber, red coral, crab's eyes, powdered hartshorn, pearls and black crab claws.[25] Such a mixture was the prized Oriental Bezoar, prescribed for different ailments. The Bezoar stone, a composition of powdered gems, received acceptance long after the magical qualities of gems were disregarded. The Bezoar stone, containing ground white amber as one ingredient, was also used as a component of the famous cordial medicine known as Gascion's powder.[26]

The famous British physician Bulleyn, cousin of Ann Boleyn, one of Henry VIII's wives, wrote the following prescription: "two drachmes of white perles; two little peeces of saphyre; jacinthe, corneline, emeraulds, granettes, of each an ounce; redded corral, amber, shaving of ivory, of each two drachmes, thin peeces of gold and sylver, of each half a scruple." Prescriptions of this type were commonly used against disease and poison.[27]

In 1624, Sir John Harrington, in his *School of Salerne,* advised: "Alwaies in your hands use either corall, or yellow amber, or some like precious stone, to be worn in a ring upon the little finger of the left hand; for, in stones, as also in herbes, there is great efficacie, and vertue; but they are not altogether perceived by us; for surely the vertue of an herb is great, but much more the vertue of a precious stone, which is very likely they are endued with occult, and hidden vertues."[28]

In 1696, the *Family Dictionary,* by Dr. W. Salmon, ordered: "For falling sickness, take half a drachm of choice amber, powder it very fine, and take it once a day in a quarter of a pint of white wine, for seven or eight days successively." If further treatment was needed, one could "take bits of amber, and in a colsestool put them upon a chafing dish of live charcoal, over which let the patient sit, and receive the fumes."[29]

Amber was also used externally as a fumigation to cure other ailments, including tonsilitis, catarrh, or running nose and eyes. It was thought that the smell of burned amber helped women in labor. In the Orient, amber fumigation was accomplished by throwing powdered amber on a hot brick. The fumes, it was thought, would strengthen the individual and give him courage from the soul of the tiger, a beast second in importance only to the dragon in Chinese mythology. "Syrup of amber," a mixture of liquid acid-of-amber and opium, was also used in China as a sedative, anodyne and antispasmodic drug. Powdered amber and oil-of-amber were prescribed as powerful diuretics.

During the eighteenth century, many criticisms of medicinal uses of stones were published. One profound critic, in an address before the Royal Society, examined bezoar-stone and the other components of Gascoin Powder and attempted to prove not only the ineffectiveness but also the negative medicinal action of the mixture. However, the enlightened critic prescribed chalk and "salt of wormwood" to replace the former remedy! In spite of these attacks, the medicinal virtues attributed to amber continued.[30] In 1770, Dr. John Cook of England, in his *The Natural History of Lac, Amber, and Myrrh; with a Plain Amount of the many excellent Virtues these three Medicinal Substances are naturally possessed of, and well adapted for the Cure of various Diseases incident to the Human Body,* wrote: "Many are the excellent virtues of amber, especially when taken inwardly in a cold state of the brain, in catarrhs, in the head-ach, sleepy and convulsive disorders; in the suppression of menses, hysterical and hypochondriacal affections; and in hemorrhages, or bleedings." Amber would "clean" any "foul inward ulcer in the lungs, kidnies or elswhere." The dose was "60 or 80 drops for grown persons, two or three times a day, in any liquid."[31] (See Fig. 4–12.)

A piece of amber placed on the nose was thought to stop excessive bleeding. It is perhaps this belief in the special blood-stilling or coagulating properties of amber that explains the use of an amber handle on the ritual Jewish circumcision knife of the eighteenth century. For some time afterward, amber was thought to be useful in connection with excessive bleeding. During the early 1900's, as medicinal practices advanced, vessels made of

THE

NATURAL HISTORY

OF

LAC, AMBER, and MYRRH;

WITH A

Plain Account of the many excellent Virtues
thefe three Medicinal Subftances are naturally
poffeffed of, and well adapted for the Cure of
various Difeafes inciden to the Human Body:

AND A

RESTORATIVE BALSAMIC TINCTURE,

which in many extraordinary Cafes gives fpeedy Relief, as are fully defcribed in the following Treatife.

In univerfum, nemo probè uti poffit medicamento com-
pofito, qui fimplicium vires priùs non accûratè didi-
cerit.

GALEN *de Comp*. lib. I.

By JOHN COOK, M.D.
Of LEIGH, in ESSEX.

LONDON:

Sold by Mr. WOODFALL, Charing-Crofs; Mr. PAR-
KER, New Bond-Street; Mr. FLEXNEY, in Hol-
bourn; Mr. DILLY, in the Poultry; and Mr. JACOB,
oppofite the Monument.

MDCCLXX.

Fig. 4-12. Title page of Dr. John Cook's *Natural History of Lac, Amber, and Myrrh*.
(1770), on the medicinal uses of amber.

compressed amber were used during blood transfusions. The amber, being a poor conductor of heat, kept a more constant blood temperature for a longer period of time than did glass or stone vessels, and thus the blood was kept from coagulating. Today, synthetic materials are used.

As late as 1896, amber was still being prescribed by physicians in France, Germany and Italy. The persistent belief that amber not only prevented infections, but acted as a charm against them, was the reason amber retained its popularity into the late 1800's and early 1900's, especially in the smoking articles industry. Its employment as a mouthpiece for cigars, cigarettes and pipes was originally talismanic. Amber was used in the Middle East for the mouthpieces of hookahs with their many hoses because of the gemstone's supposed germicidal effect.

As late as 1935, an official report of the United States Mining Bureau indicated that oil-of-amber was still used in pharmaceutical products. A volatile oil extracted from amber is used medicinally even today, in cases of infantile convulsions.[32]

ASTROLOGICAL SIGNIFICANCE OF AMBER

Most gems have been related to astrological signs and have significance for astrological readings. In ancient times, the celestial bodies were deemed to impart their powers to certain gems, which would then assist in exerting an influence on mortals wearing them. Amber, for example, was associated with the sign of Leo, because of its golden color. People born under the sign of Leo were protected by the wearing of amber, but those born under the sign of Taurus would be harmed by wearing it. It is used in reference to water signs, such as Cancer, Pisces and Scorpio, since amber is closely associated with the sea. Therefore, amber's protective qualities would be most beneficial to those born under these water signs— the Scorpios benefiting most from reddish ambers, while Cancers should select the white or water-clear varieties. Dreams which include amber signify that one will embark on a voyage in the future. Amber is specifically associated with the proper name *Anne,* and those bearing that name will be protected and kept free from illness by wearing any form of the gemstone. The ancient belief that the constellation under which one was born would influence one's fate led to the custom of wearing a stone assigned to that constellation, or a "birthstone," which served as an amulet bringing good fortune and protection to its wearer. Amber is considered by some to be the alternate birthstone for November, since its color resembles that of the topaz, the regular birthstone for that month. Amber is also the gemstone for the tenth wedding anniversary.

THE NAMES OF AMBER

The English name *amber* is based on a misconception. It is derived from the old Arabic *anbar,* a word meaning ambergris, which is the name of a curious waxy substance sometimes regurgitated in lumps from whales' stomachs. This material is in no way related to amber; although, like amber, it is occasionally washed ashore, since ambergris is also light enough to float on sea water. Ambergris has long been used, and to some extent is still used, to prevent the too-rapid evaporation of the volatile essences in perfumes. Some authorities suggest that the Arabic *anbar* originally was applied to perfumes and incenses, and because amber also was used as an incense, the word came to be applied to amber as well as to ambergris. Our present form, *amber,* is evidently derived from the Spanish *ambar* or *ambeur.* Several other languages derive similar terms from the same ancient source; for example, the French word is *ambre,* the Italian, *ambra* (or somewhat amplified as *ambra gialla*—"yellow amber") and the Portuguese, *alambre*[33]

Considerably different forms appear in the Finnish *merre-kiwa* or *merikivi,* meaning "sea-stone," in allusion to amber's marine origin. Related forms and meanings are found in the Estonian language: *meriwalk* ("sea-wax") or *meriphikaa* ("sea resin").[34] The Russian *yantar,* the Hungarian *yanta* and the Latvian *dzinters,* as well as the Old Prussian *gentar,* all appear to be derived from a common root. The Lithuanian term *gintaras,* provides a clue to their meaning, inasmuch as *ginti* in Lithuanian means "to defend," and, logically, *gintaras* means "protector" or "defender," especially when referring to a personal amulet made from amber.[35]

In Ancient Greece, the term *elektron* for amber appears in Homer's *Odyssey* and is derived from the word *elektor,* meaning "sun's glare."[36] Elsewhere in the Greek literature, the term *Elektrides* is used synonymously with "Amber Islands," or the mystic places in the Far North from which this substance came. However, to the confusion of scholars of antiquity, *elektron* had another meaning, applied to a bright yellow alloy of gold and silver. It was not until ancient Greek graves were excavated that the presence of numerous amber beads proved that this early Mediterranean civilization not only knew amber but prized it most highly; thus, the translation of *elektron* as meaning amber was verified.[37] As was mentioned earlier, it was Thales of Miletos who first drew attention to the attractive properties of amber sometime between 640 B.C. and 546 B.C., and it was from such electrostatic properties that the ancient word *elektron* came to be applied to any phenomena or device in *electronics* or *electricity.* In 1859, Franz Beckmann (see Fig. 4–13) wrote a treatise on the origin of the name in relation to these properties of amber. Curiously, the present name for amber in Greece is *beronike,* rather than *electron.*

Another name alluding to amber's power of attraction was used by the Persians. The Arabic *Kahroba* (kahruba), meaning amber, was formed by combining the two words *Kah,* meaning "straw," and *ruba* meaning "robber," thus describing amber as "straw robber" or "attractor." The modern Turkish word, *Kehruba,* is obviously related to this.

The lustrous gold-like appearance of amber has not gone unnoticed in the naming of amber. The ancient European term *glaesum* or *glesum* refers to this quality. Cornelius Tacitus, in his history, *Germania,* written about 100

Urſprung

und

Bedeutung des Bernſteinnamens

Elektron.

Von

Dr. Franz Beckmann,

Profeſſor am Königl. Lyceum Hoſianum zu Braunsberg.

Beſonderer Abbruck aus dem erſten Bande der Zeitſchrift für die
Geſchichte und Alterthumskunde Ermlands.

Braunsberg 1859.

Bei Ed. Peter Ferd. Beyer.)

Fig. 4–13. Title page of Franz Beckmann's [*Origin of the Name Elektron*] (1859).

A.D., indicates that the source of the word was from Old Germanic languages. Tacitus wrote that the people of Germania "explore the sea for amber, in their language called *glese,* and are the only people who gather that curious substance." A variant of this word is found in the Hebrew biblical writings of Ezekiel, as well as in St. John's Revelation, as *ghashmal.* These word forms all actually allude to the smooth shining yellow properties of amber. Therefore, the word *ghashmal* in ancient texts has caused scholars to argue over the accuracy of the translations, since the same term was applied to a pale gold metal of that period. It is from the old German word for amber, *glaes,* that we obtain the modern word "glass," but, strangely enough, *glaes* has not survived anywhere in its original meaning.[38]

The ability to burn is another property of amber reflected in the language of those cultures familiar with the stone. The German name *Bernstein* (Boernstein—Old German), literally means "stone that burns."* Related derivations are the Dutch *Barnsteen,* the Swedish *Barnsten,* the Scandinavian *Bernsten* and the Polish *bursztyn.* In some early Latin writings, amber is spoken of as *lapis ardens,* which also means "a stone that can be burned."[39]

In contrast to the above, the Danes call amber *rav,* while in the Frisian Islands along the coast of Denmark, it is called *rov,* and in Swedish and Norwegian, *raf.* All of these names are derived from a much older Norse word, *rafr,* a name sometimes applied to gum.[40] Some linguists suggest that this root is based on the name of the island Raunonia, mentioned by Pliny, who quoted the Roman knight sent out to search for amber and who described it as being cast ashore on Raunonia.

Arnold Buffum, in his charming study of Sicilian amber, attempts to show, through a study of *sakal,* the Egyptian word for amber, that Sicilian amber was known to the ancients. He suggests that *sakal* was not an Egyptian word at all, but was adapted from its source; that is, from *Sikeli,* a powerful race living on the east coast of the island of Trinacria (Sicily), the area where the bulk of the Sicilian amber was produced.[41] However, the Egyptian word is also strangely similar to one used in Lithuanian and Latvian languages: *saka* or *sakai,* meaning resin or gum in Lithuanian and meaning amber in Latvian. This word appears in names of places in the Baltic region, such as Sakastine (Sakaslina), or "Valley of Amber," a place located in Latvia north of Liepaja, and Sakasosta, or "port of the North," a place north of Kaliningrad.

In Dalmatia, during the Roman period, amber was called *schechel,* possibly derived from the Latin *succinum,* literally, "sap-stone," which comes from *succus,* meaning "gum." An alternate word for amber in Spanish is *succina,* representing the only survivor of the ancient Latin term in a modern language. It is also from the Latin that the minerological term *succinite* is derived.

---

* Old German forms of the word are *Burnsteyn, Boernstein* and *Bornstein.*

REFERENCES

1. Berry, E. W., "The Baltic Amber Deposit," *The Scientific Monthly* **24**: 268–278, 1927.
2. McDonald, Lucile Saunders, *Jewels and Gems*. New York: Thomas C. Crowell, 1940, pp. 156–158.
3. Ley, Willy, *Dragons in Amber*. New York: The Viking Press, 1951, p. 11.
4. Williamson, George, *The Book of Amber*. London: Ernest Benn, 1932, pp. 35–36.
5. DeBarrera, Madame, *Gems and Jewels, Their History, Geography, Chemistry and Analysis*. London: Richard Bentley, 1860.
6. McDonald, *Jewels and Gems,* p. 156.
7. Rogers, Frances and Beard, Alice, *5000 Years of Gems and Jewelry*. New York: Frederick A. Stokes, 1940, pp. 272–275.
8. Gudynas and Pinkus, *The Palanga Museum of Amber*. Vilnius: Mintis Books, 1967, p. 4.
9. Green, Roger Lancelyn, *Myths of the Norsemen*. London: The Bodley Head, 1962, pp. 75–79
10. Jobes, Gertrude, *Dictionary of Mythology, Folklore and Symbols, Part 1*. New York: The Scarecrow Press, 1961, pp. 81–82.
11. Narune, J., *Ambarita (Ambarella) Leyenda*. New York: Council of Lithuanian Women, 1961.
12. Fernie, W. T., *Precious Stones: for Curative Wear, and Other Remedial Uses*. Bristol: John Wright & Co., 1907, p. iii.
13. Leach, Marchia, *Standard Dictionary of Folklore, Mythology and Legend*. New York: Funk & Wagnalls, 1949, p. 456.
14. Rogers and Beard, *5000 Years of Gems,* pp. 272–275.
15. Fernie, *Precious Stones,* pp. 323–326.
16. Budge, E. A. Wallis, *Amulets and Talismans*. New York: University Books, 1961, p. 356.
17. Leach, *Standard Dictionary of Folklore,* p. 456.
18. Fielding, William J., *Strange Superstitions and Magical Practices*. Philadelphia: Blakiston Co., 1945, p. 56.
19. Jobes, *Dictionary of Mythology,* p. 82
20. McDonald, *Jewels and Gems,* p. 154.
21. Fernie, *Precious Stones,* p. 323.
22. Hodges, Doris M., *Healing Stones*. Iowa: Pyramid Publishers of Iowa, 1962, p. 5.
23. Villiers, Elizabeth, *The Book of Charms*. New York: Simon and Schuster, 1973. p. 50.
24. Budge, *Amulets and Talismans,* p. 356
25. McDonald, *Jewels and Gems,* p. 123.
26. Evans, Joan, *Magical Jewels of the Middle Ages and Renaissance*. New York: Dover Publications, 1976, p. 183.
27. Schweishimer, W., M.D., "The Medical Power of Jewels." *Jewelers' Circular-Keystone.* **139**, 7: 70–71, 93, 96, 1968.
28. Fernie, *Precious Stones,* p. 293.
29. *Ibid.,* p. 324.
30. Slare, Dr., "Experiments and Observations Upon Oriental and Other Bezoar Stones, 1715," in Evans, *Magical Jewels,* p. 192.
31. Cook, Hohn, M.D., *The Natural History of Lac, Amber, and Myrrh*. London, 1770, pp. 16–21.
32. Petar, Alice V., "Amber," *United States Bureau of Mines, Information* Circular 6789, 1934, p. 11.
33. "Amber," *Encyclopaedia Britannica* **I**: 718. Chicago, 1971.
34. Williamson, *The Book of Amber,* p. 50.
35. "Amber," *Encyclopedia Lituanica* **I**:84–87. Boston, 1970.
36. Ley, *Dragons in Amber,* p. 18.
37. Buffum, Arnold, *The Tears of the Heliades or Amber as a Gem*. London: Sampson Low, Marston and Co., 1897, pp. 13–14.

38. Williamson, *The Book of Amber,* pp. 50–51.
39. Keferstein, Chr., *Mineralogia Polyglotta.* Halle, 1849, p. 61.
40. Williamson, *The Book of Amber,* p. 49.
41. Buffum, *The Tears of Heliades,* pp. 13–14.

ADDITIONAL FOLKLORE REFERENCES

Kunz, George Frederick, *The Curious Lore of Precious Stones.* New York: Dover Publications, 1971, pp. 55–57.

de Cuba, Johannis, *Hortus Sanitatis, DeLapidibus. cap. ixx.* Strassburg: Jean Pryss, ca 1483–1491.

Butenas, Petras, "Gentaro Sneka," *Karys* **4:** 110–114, Balandis, 1973.

Butenas, Petras, "Gentaro Sneka," *Karys* **5:** 159–164, Geguže, 1973.

Lüschen, Hans, *Die Namen der Steine.* München: Ott Verlag, 1968, pp. 188–189.

Part 3
Scientific Aspects

5
Physical and Chemical Properties with Appropriate Tests

HARDNESS AND TOUGHNESS

Hardness is considered to be of major importance in mineral gemstones; however, it is accorded lesser importance in the case of such organic gem materials as pearl, nacre, coral, jet and amber. In all of these relatively soft substances, other considerations, such as beauty and rarity, elevate them to the rank of important gem materials. To provide a scale for determining hardness and for comparing minerals with one another, the German mineralogist, Friedrich Mohs, arranged ten minerals in order of their hardness—from the softest, designated as 1, to the hardest, designated as 10. These minerals were arbitrarily selected, and the differences between them in the continuum are not equal, but vary considerably. In spite of this, the Mohs' scale is useful, as each mineral will scratch those lower in the numerical scale.

Mohs' Scale of Hardness

1 Talc	6 Orthoclase
2 Gypsum	7 Quartz
3 Calcite	8 Topaz
4 Fluorite	9 Corundum
5 Apatite	10 Diamond

The hardness of Baltic amber varies from 2 to 2.5, depending on the type of specimen. Generally, it is slightly harder than gypsum, but can be scratched by calcite. Burmese amber and pressed amber may be as hard as 3, whereas retinites, particularly amber from the Dominican Republic, may be as soft as 1.5 to 2. Geologically younger amber tends to be softer than amber that has been buried in the ground for a longer time.

Because of amber's softness, it is mostly shaped into beads or pendants, as opposed to rings, because necklaces generally are not subjected to the rough wear rings receive. When amber "stones" are shaped for rings, they are generally cut into high domed shapes, cabochons, or even thick natural lumps, all of which provide more bulk for the piece than a faceted shape would, thus reducing the possibility of chipping.

When testing hardness, other materials than those specified in the Mohs' scale may be used to give an approximate hardness. For example, the fingernail is about 2.0 to 2.5 on the scale, and this is why one finds it extremely difficult to scratch amber with a fingernail. On the other hand, a copper penny is harder, about 3, and scratches amber, and a knife blade, about 5.5, can be used to cut amber. When a steel tool is used to scrape or cut an amber specimen, a powder or small granules are produced, resulting from amber's brittleness. Such a scrape test is useful for distinguishing amber from plastic imitations, which tend to produce small shavings rather than powder, since plastics are not as brittle as amber.

Despite amber's softness, it is tougher than most gemstones of similar hardness. On a scale of toughness from 1 to 10, amber is placed at 3.5. Baltic succinite tends to be tougher than retinite ambers. Generally speaking, Dominican amber, which is a retinite, is more brittle than Baltic amber and tends to break more easily when subjected to sharp blows. Most amber breaks with a curved, conchoidal fracture, somewhat resembling that seen on broken pieces of glass.

SPECIFIC GRAVITY

The specific gravity of amber is low, ranging from 1.05 to 1.10, or slightly higher than water.[1] However, differences may be noted even within samples taken from the same specimen. As a general rule, the clearest types or transparent ambers are more dense, whereas the osseous or "bone" amber varieties are less dense. A simple specific gravity test for amber serves to separate it from most of the usual plastic imitations. Because of amber's low density, it will float on a saturated solution of salt water, most conveniently made by adding four heaping teaspoons of salt to eight ounces of water. Though most amber specimens will float in this solution, the heavier bakelite resin, for example, with a specific gravity of 1.25 to 1.55, will sink. (However, if the sample does float, it is not necessarily definite that the piece is amber, since some imitations made of the more recent thermoplastics are lighter than amber and will even float in plain water.)

In this connection, the immersion testing fluids commonly employed by gemologists for testing gems should not be used for amber because they chemically attack and dull the surface luster. Therefore, it is best to use the saltwater solution mentioned. If more accurate determinations of specific gravity are required, the water displacement method, described in any standard gemological or mineralogical text, may be used.

OPTICAL PROPERTIES

Lacking crystalline structure, amber is *amorphous* and hence refracts light in simple fashion, like glass, or is *singly refractive.* The degree of refraction, or *refractive index,* is low, about 1.54, considerably above that of water (1.33), but far less than that of diamond (2.42). On the other hand, it is close to the index of quartz (1.55), and thus a faceted gem of amber would bend light rays to about the same degree as that found in amethyst, rock crystal and other forms of this materal.

The gemological instrument used to measure refractive index, the *refractometer,* may be used for amber. Objects such as flat-bottom cabochons or faceted beads can easily be measured, since the instrument requires that a flat, polished surface be available for contact with the sensors. Finished amber articles set in mountings, or pieces with uneven surfaces, present some difficulties in measurement. However, with a little practice, refractive index readings *can* be taken on curved surfaces, as explained in standard gemological texts.[2] A problem with using the refractometer arises from the fact that in order to test gemstones, an "optical contact" must be established between the stone being tested and the small glass prism of the refractometer. For this purpose, a special fluid is used, which, unfortunately, may dull the surface of amber if left on the piece. This problem can be eliminated if the measurement is done quickly and the specimen wiped of as soon as possible.

Another method of establishing the approximate refractive index of gemstones employs fluids, in which the test specimen is successively immersed until it seems to disappear from view, much in the manner that an ice cube seems to disappear when placed in water. A close "match" means that the index of the liquid is nearly the same as that of the specimen. Again, most immersion fluids are hydrocarbons or other organic compounds which more or less readily attack amber and therefore must be used with caution. On the whole, they should not be used, especially when amber pieces are

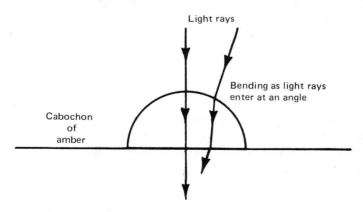

Fig. 5-1. Bending of light as it enters the stone can be measured to give the index of refraction, which helps to identify the stone.

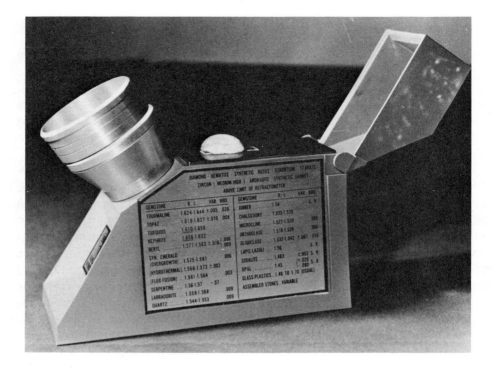

Fig. 5–2. Refractometer used for reading refractive index of gemstones.

polished, but they may be used where the effect on the surface, as in rough specimens, is not important. Needless to say, long immersion in such fluids should be avoided, since the fluids will become contaminated by the substances dissolved from the amber.

One fluid mentioned in gem literature as having a refractive index of about 1.54, similar to that of amber, and easily available from local pharmacies, is oil of clove. As a word of caution, this author tried using the fluid on both amber specimens and antique imitation amber beads, and found to her dismay that it not only dulled the surface of some of the specimens, but was difficult to remove, causing the surface of the beads to become sticky or tacky to the touch.

Polarized Light

In polarized light, obtained by "crossing" two sheets of plastic polaroid material and placing a piece of the material in question—amber, in our case—between them, transparent amber appears light when transluminated. This is what happens when amber is placed in a jeweler's polariscope with the lens set in the dark position: a rainbow effect or irridescence appears in places where internal strains are present. As the piece is rotated between the polaroids, these irridescent spots will appear as wavering bands that move across the specimen, depending on how it is turned. Though this test is not useful for discriminating amber from plastic imitations, which may be expected to behave in a similar manner, it is helpful in identification of ambroid (pressed amber). When amber is softened and pressed, the uneven stressed markings disappear and the piece develops a flowing or roiled haziness, causing it to display an evenly light appearance when rotated in the dark position of the polariscope.[3]

Fluorescence

When invisible ultraviolet light (UV) is directed upon amber, it fluoresces; that is, it emits longer wavelengths than those of UV, and thus places them in the portion of the spectrum visible to the human eye. Common fluorescent colors of amber are blue and yellow, with green, orange and white occasionally noted. Newly fractured surfaces are more fluorescent than surfaces that have been exposed to the atmosphere. The intensity of the fluorescence also varies with different varieties of amber, or fossil resins. Osseous succinite is the most highly fluorescent of all, fluorescing with a bluish white hue. According to Williamson,[4] yellow and mottled varieties of rumanite are also highly fluorescent, followed by simetite, burmite, clear-golden succinite and, lastly, Siamese amber. In this respect, Dominican amber, fluorescing blue, green or yellow, ranks with rumanite and simetite. Generally speaking, those resins with higher sulfur content are more fluorescent than are those containing less sulfur.[5]

The characteristic fluorescent hue of clear-golden Dominican specimens is most commonly blue, but yellow opaque pieces fluoresce yellow. Dominican amber also often appears to fluoresce green if viewed against a light source (back lighting), but this may result from selective absorption of light rather than from true fluorescence. Some varieties of Dominican amber and Sicilian amber not only fluoresce under a black light (UV), but also in reflected light. Dominican amber pieces possessing this quality appear as though a blue film covers the surface of the amber when ordinary light strikes it. Transparent pieces of Burmese amber usually fluoresce blue, but the osseous (or "root") amber from Burma may fluoresce in the orange range.[6]

DIAPHANEITY AND STRUCTURE

As mentioned previously, the differences in diaphaneity or turbidity in amber specimens is caused by the enclosure of vast numbers of microscopic bubbles of varying sizes. In those areas which are cloudy, the bubbles are too small to be seen with simple lenses, but thin sections under a high-power microscope reveal minute voids distributed throughout otherwise pure clear resin. Photographs made using a scanning electron microscope show the distribution, size and shape of the pores in various specimens. These photographs indicate that the smaller the pores, the greater the quantity. It

Table 5-1. Bubbles influencing diaphaneity.

Type of Amber	Number of Bubbles (per square millimeter)	Size of Bubbles (millimeters in diameter)
Bone (osseous)	900,000	0.0008 to 0.004 mm
Bastard	2500	0.0025 to 0.012 mm
Fatty (flohmig)	600	0.02 and larger

is the number and size of these cavities that produce the different appearance in amber varieties.

The early work of Max Bauer provides a count of the bubble enclosures in the three most prevalent amber types (see Table 5.1). Bauer reports that the bubbles are smallest and most numerous in osseous amber, "having a diameter of from 0.0008 to 0.004 millimeter and a distribution of 900,000 per square millimeter. Flohmig amber contains the smallest number of bubbles, 600 per square millimeter, but these, with a diameter of 0.02 millimeter, have the maximum size."[7] When amber contains no bubbles, it is perfectly transparent.[8] Cavities in the cloudy amber can be filled with a transparent substance with the same refractive index as the surrounding mass, causing the turbidity to disappear, and rendering the specimen perfectly transparent. (This is sometimes accomplished artificially, as explained in the chapter on preparation and working of amber.)

The presence of air bubbles, of course, makes amber very porous. The size and density of the pores influence the structure of the amber, which can be differentiated into three types:

1. *Compact.* Slightly porous, heavy.
2. *Nodular.* Moderately porous.
3. *Foamy.* Strongly porous, light.

The above structural classifications are presently used by the Polish Academy of Science in Warsaw to sort various types of amber for displays in the Amber Museum of Warsaw.

COLOR VERSUS RESIN SOURCE

Early studies of the coloration of amber suggested that the microscopic bubble inclusions were the sole cause of the color variations. However, current studies relating fossil resins to recent resins suggest that certain colors are characteristic of certain tree sources. For example, recent pine trees, even those living today, produce resins in golden yellows from transparent to opaque, similar to those found in the most common amber specimens. Such trees also produce white, ivory-colored and occasionally the rare blue variety of resin. The greenish shades in amber closely resemble resin exuded from a recent spruce tree. The Polish Museum of Science theorizes that reddish tints are likely in resins of deciduous trees, such as the cherry and plum species, whose modern representatives exude a resin with a cherry red hue.

Amber from the Dominican Republic, which often possesses a reddish tint, is thought to be related to a leguminous tree.[9]

SOLUBILITY

Amber is insoluble in water and only slightly soluble in ether, and then only after long duration of contact. When a drop of ether is placed on amber, the ether evaporates before the amber has time to soften. Copal and other recent resins are more quickly attacked by ether and become "sticky," thus providing a convenient test for differentiating amber, or fossilized resins, from recent resins such as copal. Crushed particles of amber are soluble in sulfuric acid, as shown in Table 5–2. Because amber is slightly soluble in hydrocarbon compounds, contact with methylene iodide, biomoform or acetylene tetrabromide—heavy fluids sometimes used by gemologists—should be avoided.

Table 5–2. Solubility of Amber.

Solvent	Clear Golden Amber	Bone or Osseous Amber
Ether	18 to 23 percent	16 to 20 percent
Alcohol	20 to 25 percent	17 to 22 percent
Turpentine	25 percent	
Chloroform	26 percent	
Benzene	trace	
Concentrated sulfuric acid	soluble if crushed	
Hot concentrated nitric acid	soluble	

REACTIONS TO HEAT

Amber is a poor conductor of heat; therefore, it feels warm to the touch, in contrast to many other gemstones, which feel cool. When rubbed briskly with a cloth or gently heated, a typical "piney" odor is emitted. With further heating, amber softens at about 150 to 180 degrees centigrade, then swells and emits volatiles. Between 250 and 375 degrees centigrade, it decomposes, evolving white fumes with a strong pine odor, and releasing succinic acid and oil-of-amber. Amber ignites and burns readily, producing a bright yellow flame streaked with green and blue flashes.

When crushed Baltic amber is introduced into a test tube and heated, a frost-like sublimate of succinic acid crystal appears on the upper, colder portions of the tube. Filter paper moistened with a solution of lead acetate turns black when held in the tube, as a result of the presence of hydrogen sulfide in the fumes.

The residue left after volatile substances have been driven off by heating the amber to 280 degrees centigrade is termed *colophony* and consists principally of succinic acid and oil-of-amber, which, when dissolved in turpentine and linseed oil, provides the basis for the characteristic dark-hued amber varnish extensively used in the last century.

ELECTRICAL PROPERTIES

Amber becomes strongly charged with negative electricity, attracting small bits of paper, lint or straw when rubbed on fur or velvet. This property of producing static electricity has been suggested as a test for identifying true amber in the past, but it is no longer useful since recent plastics possess similar qualities. Vigorous rubbing heats the amber and one may detect the faint odor of pine after doing so, but the amber remains smooth and untouched by the rubbing. On the other hand, such rubbing of non-fossilized resins as copal often results in the heat softening the surface layers to the point where they become sticky, providing a convenient clue to identity.

MINERALOGICAL CLASSIFICATION

Because amber was found in the ground, it seemed logical in the past that amber should be considered a mineral and classified as such among other organic materials as bitumens, mineral waxes and coals. A commonly accepted classification scheme by H. Strunz[10] is as follows: 1. elements; 2. sulfides; 3. halides; 4. oxides and hydroxides; 5. nitrates, carbonates, borates; 6. sulfates; 7. phosphates, arsenates, vanadates; 8. silicates; and 9. organic substances, the last including amber and other fossil resins. Another classification scheme, in use by the Polish Academy of Sciences Museum of the Earth, places such organic materials into a group called *caustobioliths,* a term that states these to be combustible materials, of biogenic origin, which resemble stone. Amber is included in a sub-group, along with other resins and mineral waxes, as *liptobioliths,* meaning further that the materials of which they were composed are resistant to the decaying process and do not mold, and thus are the same constituents of the original plant.[11]

CHEMISTRY

The broad composition of amber varies from specimen to specimen but is generally 79 percent carbon, 10.5 percent hydrogen and 10.5 percent oxygen. In the amber from the Carpathian Mountains, such as rumanite and other varieties, as well as amber from several other localities, small quantities of sulfur occur. Some types of amber may occasionally contain inclusions of inorganic materials such as calcite or pyrite.

Various investigators in the past have assigned chemical formulas to amber—$C_{10}H_{16}O$,[12] $C_{12}H_{20}O$,[13] $C_{40}H_{64}O$[14]—all of which are without real significance inasmuch as amber is not a single organic compound, but rather, a complex mixture of several compounds in varying proportions. More significantly, amber is composed of *isoprenoids,* a particular type of hydrocarbon compound found in natural resins. These isoprene units (C_5H_8) link together, forming more complex organic molecules called *terpenoid* compounds. These are characteristically found in the sap of plants, with the proportions of the isoprene units varying from one plant species to another. Recent work by botonists has established the presence of the isoprene units, not only in ambers, but also in living plants. By the resemblances in the quantities of these compounds, tentative identifications can be made of the ancient species of amber-producing trees, giving some insight into the evolution of the present day resin-producing vegetation.

A significant constituent of Baltic amber is succinic acid, whose formula may be written as $COOH(CH_2)_2COOH$[15] or as an isoprene structure CH_2CO_2H.[16] In general, the weight percent of succinic acid in Baltic amber

|
CH_2CO_2H

ranges from 3 to 8 and varies according to type. Clear, transparent specimens contain the least amount, approximately 3 percent to 4 percent, whereas cloudy specimens generally contain more. Frothy or foam-like ambers contain the most—sometimes as much as 8 percent. Baltic succinite contains a considerably higher portion of succinic acid than do other fossil resins, whether retinite or succinite.

To determine whether an amber specimen is succinite rather than retinite, a chemical analysis must show that it contains succinic acid. Succinite, or amber containing succinic acid, is produced in other deposits in various localities besides the Baltic, though this formation is by far the largest. Therefore, the presence of succinic acid in an amber specimen is usually considered to be a clue that the amber most likely originated in the Baltic area. In archaeological research where amber artifacts are found, the find is tested for succinic acid to determine the place of origin. If the specimen does not contain succinic acid, the origin is considered to be other than the Baltic formation.

Testing for succinic acid is not a task for the amateur, since chemical analysis requires destructive pyrolysis or heating to decompose the amber. The specimen is ground, then heated, with close observation made of the ''melting'' process. After volatile substances are driven off, the residue is boiled in ether, then dried and weighed. After weighing, more ether is added, along with nitric acid. Deposits of crystals are examined for the presence of succinic acid. After settling for 12 hours, the remaining liquid is mixed with distilled water and separated into fractions, one of which is heated and smelled for turpentine or resinous odors, and another is filtered and diluted with sulfuric acid to indicate the presence of succinic acid deposits.[17]

A modern technique for distinguishing succinite from retinite is to employ an infrared (IR) spectrometer to record the IR spectra and compare spectral characteristics of the resins tested with those of known specimens. Variations in the spectra are indicative of specific differences in molecular structure of various types of resin, each species having its own characteristic spectral pattern. X-ray diffraction studies also have been used to examine amber resin to detect the presence of crystalline components. Though amber itself is amorphous (without crystalline structure), crystals of succinic acid may be detected within the amber mass.

REFERENCES

1. Bauer, J., *Minerals, Rocks and Precious Stones.* London: Octopus Books, 1974, p. 552 (gives the specific gravity as 1.08 to 1.1). Sinkankas, J., *Gemstones in North America.* New York: Van Nostrand, 1959 (stresses considerable variation in specific gravity, giving the range as 1.05 to 1.1, but cautions pieces can be found as low as 1.031 and as high as 1.168).
2. For instructions in how to use a refractometer, see: Webster, R., *Gems, Their Sources, Descriptions, and Identifications.* Hamden, Connecticut: Archon Books, 1975, or London: Butterworths, 1962, 1970; Liddicoat, R. T., Jr., *Handbook of Gem Identification.* Los Angeles: Gemological Institute of America, various editions.
3. Liddicoat, R. T., Jr., *Handbook of Gem Identification.* Los Angeles: Gemological Institute of America, 1969, p. 327.
4. Williamson, G., *The Book of Amber.* London: Ernest Benn, 1932, p. 94.
5. Strong, D. E., *Catalogue of the Carved Amber in the Department of Greek and Roman Antiquities.* London: The Trustees of the British Museum, 1966, p. 7.
6. Webster, *Gems,* pp. 439–445.
7. Bauer, Max, *Precious Stones* II: 535–557. London: Charles Griffin and Co., 1904 (reprinted by Dover Publications, 1968).
8. Ley, Willy, *Dragons in Amber.* New York: The Viking Press, 1951, p. 37.
9. Zalewska, Zofia, *Amber in Poland, A Guide to the Exhibition.* Warsaw: Wydawnictwa Geologiczne, 1974, p. 27.
10. Strunz, H., *Mineralogische Tabellen.* Leipzig: Akademische Verlagsgesellschaft Geest und Portig K., 1966.
11. Zalewska, *Amber in Poland,* p. 35.
12. *Ibid.*
13. Bauer, J., *Minerals, Rocks and Precious Stones,* p. 552.
14. Dana, Edward S., *Dana's Textbook of Mineralogy, 4th ed.* New York: John Wiley & Sons, 1949, p. 776.
15. Sinkankas, *Gemstones,* p. 597.
16. "Amber Substance of the Sun," *Chemistry* **45**: 21–22, April 1972.
17. Allen, Jamey, "Amber and Its Substitutes, Part II, Mineral Analysis," *The Bead Journal* **2,** *4:* 11, Spring 1976.

ADDITIONAL REFERENCES

Alexander, A. E., "Renewed Interest in Amber, Organic Compound, Cited, The Gemologist's Corner," *The National Jeweler,* March 1976, p. 27.

Anderson, B., *Gemstones for Everyman.* New York: Van Nostrand Reinhold, 1971, pp. 303–305.

Anderson, B., *Gem Testing.* New York: Emerson Books, 1948, pp. 210–214.

Arem, Joel, *Color Encyclopedia of Gemstones.* New York: Van Nostrand Reinhold, 1977.

Arem, Joel, *Gems and Jewelry.* New York: Bantam Books, 1975, pp. 91–93.

Parsons, Charles J., *Practical Gem Knowledge for the Amateur.* San Diego: Lapidary Journal, 1969, pp. 136–137.

Rice, Patty, "Amber is Back," *Gems and Minerals* **488**: 14–17, 1978.

Sinkankas, John, *Van Nostrand's Standard Catalog of Gems.* New York: Van Nostrand Reinhold, 1968, p. 54.

6
SCIENTIFIC STUDIES
OF AMBER

If thou couldst but speak, little fly,
How much more would we know about the past?
Kant

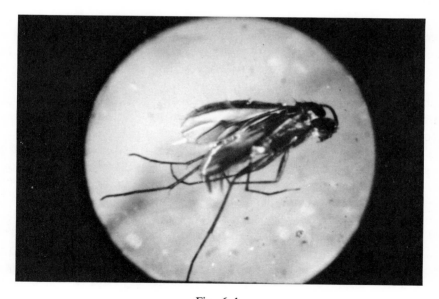

Fig. 6–1.

HISTORICAL SPECULATIONS ON THE ORIGIN OF AMBER

Amber has interested scientists for centuries. The mystery of its origin stimulated inquiry and speculation by Sophocles (490 B.C. to 406 B.C.) and Aeschylus (525 B.C. to 456 B.C.). Ancient myths associated amber with tears, apparently because of the specific, drop-like form of the pieces, as in the myth of Phaëthon's sisters shedding tears of amber from the poplar trees

145

to which they were transformed. Amazingly, even this ancient tale hints at the vegetable origin of amber.

Pliny's *Natural History,* written about 79 A.D., presented a series of fantastic views about the origin of amber as expressed by his predecessors. He dismissed such views as groundless and stated emphatically that amber was of vegetable origin and formed from a coniferous tree gum which hardened with time and fell into the waves of the sea, later to be deposited on the shore. As proof, he pointed to such properties of amber as its capacity to give off the odor of resin upon rubbing, the smoky flame resembling that of pine resin's when burned, and the presence of insects within amber.

Strangely, this essentially accurate information on the origin of amber was ignored or forgotten, and it was not until the Renaissance evoked a new interest in amber that the question was re-examined. Three outstanding scientific authorities of that time all concluded that amber was of inorganic origin; namely, Athanasius Kircher, S.J. Philippus Aureolus Bombastus Theophrastus von Hohenheim (who called himself "Paracelsus") and Georg Bauer (who called himself "Agricola"). The last-named, Georgius Agricola (1494 to 1555), the leading authority on mining and processing of useful minerals, believed amber was formed by solidification in air of liquid non-organic bituminous matter flowing from the "bowels of the earth." In fact, he thought petroleum was merely liquid amber, and, as proof, he noted that amber often contained liquid-filled bubbles or uncongealed "petroleum." He believed that petroleum, as it seeped from the bottom of the Baltic, rose to the surface in liquid form and embedded small gnats and flies at the surface.

Adherents of the "petroleum theory" lived in Italy, Switzerland and southwestern Germany and never traveled to the Baltic to see amber as it was found. Yet they described oil slicks on the Baltic Sea and told of amber being washed upon the shore while still in a tar-like form. Some of them even speculated that oil was seeping out of the sands of Samland's coast. However, no Baltic fisherman had ever seen oil slicks or found an amber lump in a tar-like condition.[1]

Athanasius Kircher (1601 to 1680) supported the liquid petroleum theory based on the find of small fish imbedded in amber specimens in his collection, and suggested that they were trapped in the "oil" as it rose to the surface. Unfortunately, since about 1500 A.D., fake fish, lizard and frog inclusions were placed in amber specimens by hollowing out the piece, inserting the small animal, filling the interstices with rapeseed oil and resealing. No fish has ever been found naturally imbedded in amber.[2] In 1742, Nathanael Sendel (Nathanaele Sendelio)[3] published the first volume exposing these fakes in amber.

The belief in amber's inorganic origin predominated for several hundred years, but other, equally unfounded theories appeared which attributed the origin of amber to living creatures. For example, some authorities thought amber was the congealed honey of wild bees, and others believed it to be a secretion of ants. Still others were sure it was excreted by whales, thus con-

HISTORIA
SVCCINORVM

CORPORA ALIENA

INVOLVENTIVM

ET

NATVRAE OPERE

PICTORVM ET CAELATORVM

EX

REGIIS
AVGVSTORVM

CIMELIIS

DRESDAE CONDITIS

AERI INSCVLPTORVM

CONSCRIPTA

A

NATHANAELE SENDELIO D.

MEDICO REGIO ET PHYSICO

ELBINGENSI ORDINARIO.

Fig. 6–2. Title page of Nathanaele Sendelio's *Historia Succinorum* (1742).

fusing amber with ambergris, a regurgitation of whales, even though at that time the distinction between these two products was already well known.

About 1750, George Louis Leclerc, Comte de Buffon, the eminent French naturalist, proved that coal and some other "minerals," including petro-

Fig. 6-3. First illustrations of fakes in amber by Nathanael Sandel (1742).

leum, were of organic or vegetable origin. Buffon asserted that they were re-mains of old forests, and if this were true, the amber, even if it were "solidified petroleum," also would be an organic product.

About the same time, Karl von Linne (Carolus Linnaeus) the Swedish naturalist and founder of modern systematic botany, collected facts about amber which convinced him it was vegetable in origin. By now, the assembled knowledge of the nature of beds of amber, the mineral remains found in it, and the pieces of amber found in chinks and under the bark of fossilized trees forced the re-examination of Agricola's petroleum theory.

Mikhail Vassilievitch Lomonosov, the Russian scientist, was the first to denounce the petroleum theory in its entirety. While compiling the catalog of the Mineral Collection of the Russian Academy of Sciences in 1741, Lomonosov closely studied amber and found evidences confirming amber's vegetable origin. Emphasizing its unique physical and chemical properties, he noted the nearness of the specific gravity of amber to that of pine resin and pointed out that "water separated from amber by chemical means smells of succinic acid, a property of growing things."[4] In a speech delivered in 1757 to the Academy in St. Petersburg, he denounced the petroleum theory and stated that amber had to be the fossilized resin of some ancient tree and that the prevailing explanation of the presence of insects in amber was unfounded.

Ten years after Lomonosov's speech, a great expert on amber from Königsberg, Friedrich Samuel Bock,[5] confirmed Lomonosov's findings in a monograph whose title is freely translated as "Attempt at a Short Natural History of Prussian Amber." Bock said that everything found naturally imbedded in amber related to a forest environment; for example, the flies, gnats and ants, all of which crawl along the trunks of pine trees. He also noted that amber was found deep in the soil associated with a blue loam. Based on this observation, he reasoned that during Pliny's time, there may have existed islands between Sweden and Samland covered with forests which produced resin, but that these islands had sunk during the intervening centuries. It was from these forests, Bock said, that amber was continually being washed ashore, especially during storms that churned the sea water.

Elaborating on the studies and logic of Bock, scientists began placing the time of the existence of the amber forest farther and farther into the past. Buffon became convinced that the origin of the earth preceded recorded history by about 80,000 years, and that plants and animals during this

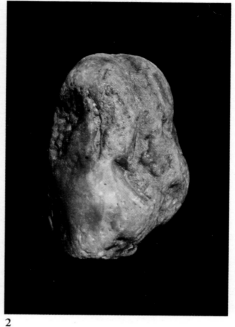

1

2

3

1. Necklace of "sea–stone" mottled "bastard" amber strung with silver spring spacers. Nodules range from 25 mm to small tear-drop forms. 36 inches long. Lithuanian style, Author's Collection.

2. Pit amber lump showing outer crust. "Bastard" to foamy amber. Size: 4 in. x 3 in., 70 g. Baltic amber, Author's Collection.

3. Amber varies within the yellow range. Natural Baltic amber carved scarabs from Poland. Approx. 25.4 mm x 12.7 mm. Amber Forever Collection, Florida.

4

5

6

4. Faceted beads of transparent amber darkened from exposure in air over a period of years. Some of the beads have insects enshrined within. Circa 1900 to 1930, Baltic amber, Germany. A. Calomeni and Author's Collection.

5. Exquisitely faceted beads of clear golden amber and reddish-brown amber. Circa early 1900's, Baltic amber, Germany. Cloudy yellow amber cigarette case, incrustation technique. Kuhn's Baltic Amber Collection, Florida.

6. Clear, transparent Baltic amber of various shades ranging from lemon yellow to orange to brown. Author's Collection.

7

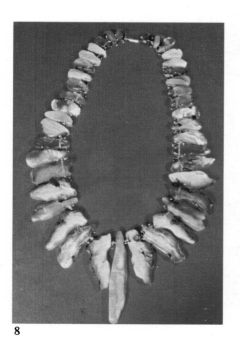

8

9

7. Fatty amber beads. The largest bead is approximately 25.4 mm in diameter. Necklace contains 45 round beads and is 32 inches long. Lithuania, N.J. Vaznelis Collection, Gifts International, Inc., Chicago.

8. Necklace contains polished amber slabs which retain the outer crust. Pieces range from transparent to fatty to "bastard" amber. Longest slab is approximately 64 mm. 96 g. Each slab is separated with a silver twisted wire spacer. Lithuania, Author's Collection.

9. Cloudy semi–bastard amber pendant, sometimes called "egg yolk" or "custard" amber, due to the swirl effect caused by mixing of transparent and opaque amber. 17.3 g. Poland, Author's Collection.

10

11

12

10. Mosaic made of different varieties of amber including rare blue amber. Lithuania, Courtesy Palanga Museum of Amber, Palanga, Lithuania, S.S.R.

11. Cloudy amber bracelet with several pieces of rare blue amber. Lithuania, N.J. Vaznelis Collection, Gifts International, Inc., Chicago.

12. Semi–bastard amber, oblate shaped beads with free form pendant of fatty to "bastard" amber. 20 in. in length, graduated from 17.0 mm at center to 10.0 mm diameter at ends. Pendant is approximately 50.0 mm in length by 40.0 mm in width and 20 mm in thickness. 94 g. Poland, Author's Collection. Photo by Pat Hall.

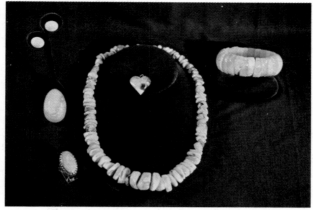

13

14

15

13. Amber earrings, ring, bracelet, necklace and amber egg made of semi–bastard amber, ranging in hue from yellow to white. Called "egg yolk" amber. Poland, A. Calomeni and Author's Collection.

14. "Tomato" amber necklace, bracelet and earrings, ranging in color from cream to orange–brown osseous amber. Necklace contains 48 tumbled stones; bracelet 13; earrings one each. Baltic, A. Calomeni Collection.

15. Oval shaped beads of white bone or osseous amber. The necklace is 39 inches in length with 37 amber beads graduated from 21.0 mm at center to 9 mm at ends. Poland, Author's Collection.

16

17

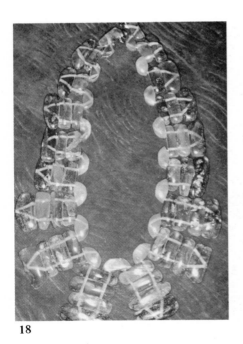

18

19

16. Rare green amber necklace made of free form tumble-polished lumps, graduated from 40.6 mm to 25.0 mm, spaced with lemon colored amber rhondels, 34 inches in length. 143.4 g. Large green amber pendant, 70.0 mm by 100.0 mm. Baltic, A. Calomeni and Author's Collection.

17. Rare black and white amber necklace, pendant and ring. White osseous amber contains black speckles resulting from fossilized pine needle inclusions. Tumble polished stones and free form pendant. Necklace is 30 inches in length. Baltic, Poland, Author's Collection.

18. Unusual Lithuanian style necklace of transparent to fatty amber, 70 pieces. N. J. Vaznelis Collection, Gifts International, Inc., Chicago.

19. Transparent amber with etched design on the under surface, similar to the technique called verre églomisée used extensively in the 17th century amber art. M. Kizis Collection, Canada.

20

21

22

20. Amber chess set constructed using natural color variations within the amber. The board is made using the incrustation technique. N. J. Vaznelis Collection, Gifts International, Inc., Chicago.

21. Carving of an amber Lord or Teutonic Knight. Made of natural "bastard" amber, 77.0 mm by 50.0 mm approx. Baltic, Poland, Amber Forever Collection, Florida.

22. Large lump of natural transparent amber which shows botanical inclusions with back lighting, 4½ inches by 5½ inches, 125 g. Baltic, Author's Collection.

23

24

25

26

23. Lithuanian amber mosaic. The scene on the black wooden background is made entirely of natural Baltic amber. This mosaic depicts a chapel on the Baltic coast near Palanga. Some of the pieces are cut and polished while other lumps are left unpolished. (10½ inches by 21 inches). Author's Collection.

24. Sculpture of growing mushrooms made using both transparent and cloudy amber. Polished and unpolished pieces are combined to effectively utilize the various textures of amber. Lithuania, Author's Collection.

25. A crocheted collar or bib made of all varieties of natural Baltic amber. Each polished tear–drop shaped amber piece is crocheted directly into the bib similar to the way amber was sewn to clothing during ancient times. 238 g. Author's Collection.

26. Necklace of osseous amber styled in Lithuania. Silver looped wires connect each amber piece. The largest center stone is approximately 29.0 mm in length. N.J. Vaznelis Collection, Gifts International, Inc., Chicago.

27

28

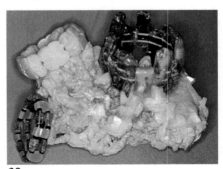

30

29

27. White amber is used to form a mosaic scene depicting the marshes near Lietuva, Lithuania (12 inches by 16 inches). V. Tomaslunis Collection.

28. Sculpture using natural amber to form a cross. (3½ inches by 2½ inches). Lithuania, N. J. Vaznelis Collection, Gifts International, Inc., Chicago.

29. Amber and silver sculpture by famous amber artist, B. Mikulevicius. N. J. Vaznelis Collection, Gifts International, Inc., Chicago.

30. Amber bracelets from Lithuania. Top left, cloudy yellow amber; top right, natural transparent amber; bottom left clarified and pressed amber pieces. N. J. Vaznelis Collection, Gifts International, Inc., Chicago.

31

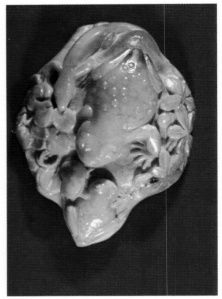

32

33

34

31. Carved rose made from fatty amber (2 inches by 1½ inches), West Germany. Amber Forever Collection, Florida.

32. Large cloudy "bastard" amber lump carved into a frog surrounded by plants, (approx. 4 inches by 5 inches), West Germany. Kuhn's Baltic Amber Collection, Florida.

33. Baltic Sea scene made in amber mosaic (8 inches by 12 inches), Lithuania. V. Tomaslunas Collection.

34. Amber mosaic depicting the folktale of Lithuania called "Torch of Happiness." V. Tomaslunas Collection.

35

36

37

35. Amber lump from the Caspian Sea. The pale champagne color lump is transparent and contains prehistoric ants and wood boring beetles. Author's Collection.

36. Dominican Republic amber styled by native craftsmen. Pieces range from cloudy yellow to transparent browns. Author's Collection.

37. Variations in colors showing "blue" and red Dominican amber. Courtesy Giovanni Creations, San Juan, Puerto Rico.

38

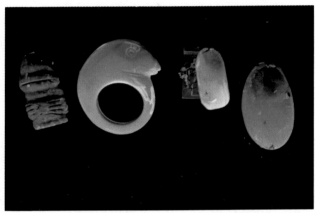

39

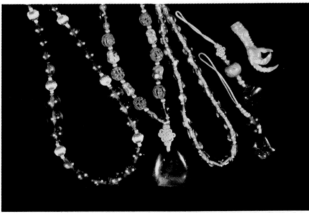

40

38. Dominican amber contains many insect inclusions and may "fluoresce" green with backlighting. Author's Collection. Photo by Pat Hall.

39. Dominican amber is highly fluorescent under ultraviolet light. Picture was taken under long wave U.V. light with Wratten 2A filter F 2.8/ 3 min. Author's Collection.

40. Amber ornaments from China. Left to right. Mandarin style beads from early 1900's made of safety celluloid imitations. Burmese amber necklace with pendant of transparent amber mottled with dark colored swirls, called "tiger amber," flat discoidal beads of the same material, "root" amber beads alternating with carved apricot pits. A necklace of transparent lemon yellow Siamese amber. Red Burmese amber drops. Carved claw shaped watch fob made of horn imitation amber. Author's Collection.

41

43

42

44

41. Chinese style necklace with round Burmese amber beads and carved jadite. Ceremonial style. China. Amber Forever Collection, Florida.
42. "Young amber" from Tanzania, Africa. Pleistocene age. This semi–fossil resin is pale transparent color and softer than Baltic amber.
43. Necklace made of unusually large free form lumps of transparent Baltic amber with "sun–spangle" inclusions. Lumps graduate from one large center piece at 50.0 mm in length, six pieces are approx. 40.0 mm, two pieces 35.0 mm, two, 30.0 mm. 129.4 g. Poland, Author's Collection.
44. Baltic amber bracelet, golden transparent amber with "sun–spangles" or lily pad shaped inclusions. West Germany, Author's Collection.

45

46

47

48

45. Reconstructed amber from Almazjuvelirexport, Moscow, Russia. Note undulating lines in the amber's appearance. H. Evens Collection.

46. Pressed amber from Almazjuvelirexport, Russia. Note the evenly matched color. Author's Collection. Photo by Pat Hall.

47. Dark brown pressed amber beads in the photo still show the hazy form of small chips of natural amber. These were possibly formed by the Spiller pressed amber technique. The yellow amber in the photo is natural tumble-polished amber. Author's Collection.

48. Pressed amber bracelets. The color variations result from different temperatures and duration of heating used during the process of manufacturing. Gifts International, Inc., Chicago.

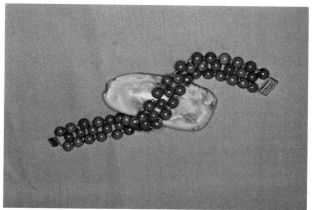

49

50

51

52

49. "Antique" pressed amber bracelet, circa 1890 to 1900. Green dye was often added during the manufacturing process of pressed amber. The yellow lump is natural Baltic amber. A. Calomeni Collection.

50. Sculpture of mushrooms made of natural amber and heat treated Baltic amber. M. Kizis Collection, Canada.

51. "African Amber" or imitation copal. The imitation beads often show flow lines indicating the material was formed in long rods, then cut into beads. Some are made of synthetic resins lighter than amber and will float in salt water. Courtesy Beada–Beada, Royal Oak, Michigan.

52. Bakelite plastic imitations, circa early 1900's are the most common amber imitations and are often unknowingly sold as genuine amber in estate sales and antique shops. A. Calomeni and Author's Collection.

53

54

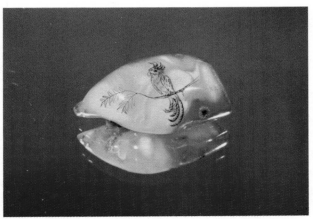

55

53. Imitation amber produced by Slocum Laboratories in Royal Oak, Michigan. The imitation contains insect inclusions and discoidal petal–like inclusions. It is made in red and orange hues. Miner's Den, Royal Oak, Michigan.

54. Cloudy and osseous Baltic amber provides an excellent base for scrimshaw art work. These pieces were hand polished by the author and scrimshaw art was applied by Bryce Barker, Michigan scrimshaw artist. Author's Collection.

55. Cloudy Baltic amber brooch with scrimshaw art by Bryce Barker. Author's Collection.

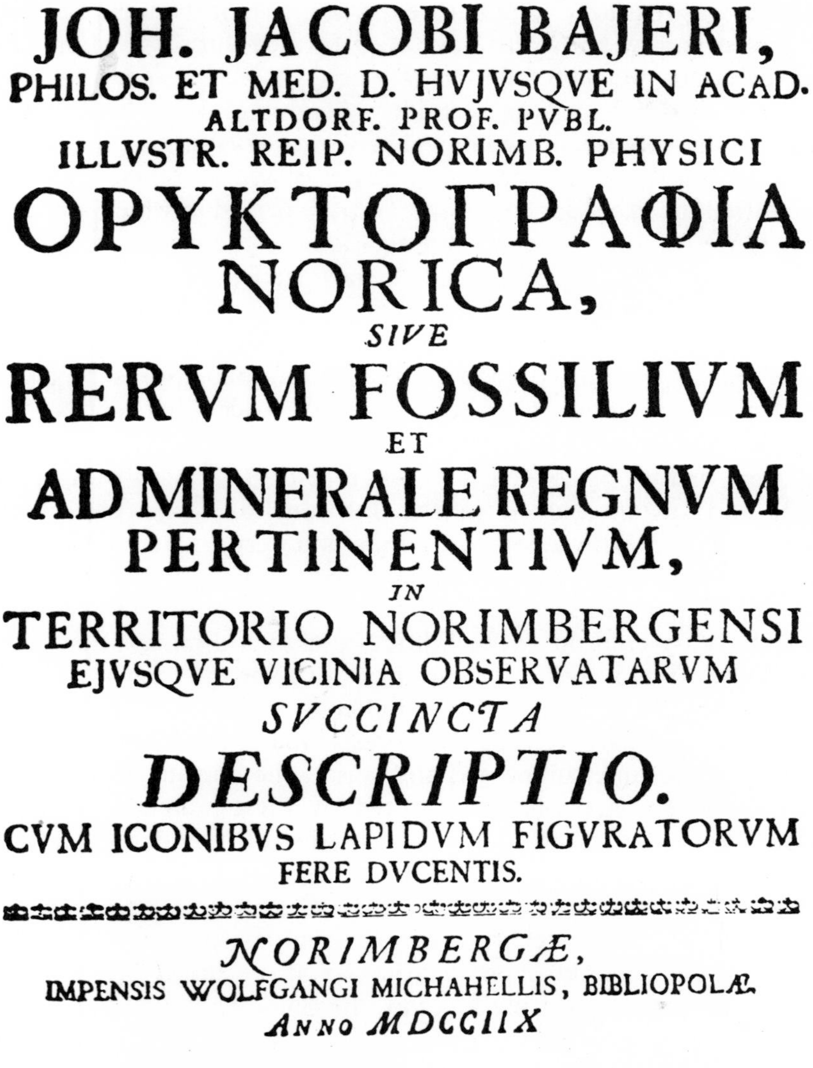

JOH. JACOBI BAJERI,
PHILOS. ET MED. D. HVJVSQVE IN ACAD.
ALTDORF. PROF. PVBL.
ILLVSTR. REIP. NORIMB. PHYSICI

ΟΡΥΚΤΟΓΡΑΦΙΑ
NORICA,
SIVE
RERVM FOSSILIVM
ET
AD MINERALE REGNVM
PERTINENTIVM,
IN
TERRITORIO NORIMBERGENSI
EJVSQVE VICINIA OBSERVATARVM
SVCCINCTA
DESCRIPTIO.
CVM ICONIBVS LAPIDVM FIGVRATORVM
FERE DVCENTIS.

NORIMBERGÆ,
IMPENSIS WOLFGANGI MICHAHELLIS, BIBLIOPOLÆ.
ANNO MDCCIIX

Fig. 6–4. Title page of Bajeri, writing in the 1700's on amber.

period grew in places where they no longer exist, with some species now extinct. By 1811, there was no longer any doubt amber was the product of pine-like trees, but because no living tree could be found producing amber, amber had to be resin from a prehistoric tree. Today, this is known to be true, but the antiquity of the trees in the amber forest has been more accurately determined.

During the 1830's, geologists began studies of deposits in which amber was found, discovering fossilized plants and leaf prints in lignite coals dated to the Tertiary period. The expert on Tertiary fossils, Oswald Heer,[6] of Switzerland, identified remains of swamp cypress and sequoia similar to North American varieties, but it was not until 1850 that G. Zaddach[7] found that amber was fossilized long before the first tree of the lignite forest grew. Baltic amber is now known to have been formed during the Early Tertiary period, or about 40 to 60 million years ago.

Since amber occurs only in secondary deposits, having been transported from its original source by geological processes resulting in changes in land formations, glaciation and changing action of sea currents, scientists have not identified the exact epochs of the Tertiary period in which amber forests existed. However, stratigraphic investigations of the present amber-bearing deposits of the Baltic region show that they are Upper Eocene. This means that the primitive forest must have come into existence no later than 40 million years ago, and may have existed before then, but the exact time span of these forests has not been definitely established. Some of the richest Baltic deposits of amber also occur in the Lower Oligocene strata, suggesting that the forests still existed during that epoch.

The areas of Europe covered by these forests are still only approximately defined, and, as mentioned in the Introduction, extensive continental changes took place between the Cambrian and Tertiary periods in the northeastern region of Eurasia, or the area called pre-Fennoscandia. During the Paleocene, or the earliest epoch of the Tertiary, pre-Fennoscandia is believed to have covered the entire Baltic Sea area and more.[8] Early scientists of the 1800's offered several descriptions of the boundaries of this primeval forest, but these earliest descriptions limited its size, with G. Berendt[9] believing, as Bock had earlier, that the forests had existed on an island near Sweden which became submerged and was at the bottom of the Baltic Sea. Berendt theorized that amber deposited by these sunken forests washed upon the Baltic shore from Klaipeda to Gdańsk. It was Berendt who placed the location of the forest at latitude 55 degrees and longitude 19 to 20 degrees.

By the end of the nineteenth century, scientists such as 0. Heer[10] (1869) and H. Conwentz[11] (1890) expanded the size of the amber-bearing forests to cover the area now comprising Finland, the Baltic Sea and large portions of Scandinavia, as well as northern regions of Latvia, Lithuania, Poland and Germany.

CURRENT GEOLOGICAL THEORIES ON THE ORIGIN OF AMBER

Today, it is believed that the amber-bearing forest extended far beyond the boundaries of pre-Fennoscandia, covering much greater regions than imagined by early authors. In 1943, W. L. Komarow[12] described it as a vast wooded area covering much of the northern Eurasian continent, reaching almost as far south as the Black Sea. He included the area from the Scandinavian mountains to the northern Urals, the northern regions of Pomerania, Lithuania, Latvia, portions of Byelorussia and the Ukraine. (See Fig. 6–6.)

Using radioactive methods for dating rocks, scientists now calculate that the Upper Eocene epoch began approximately 40 million years ago, while

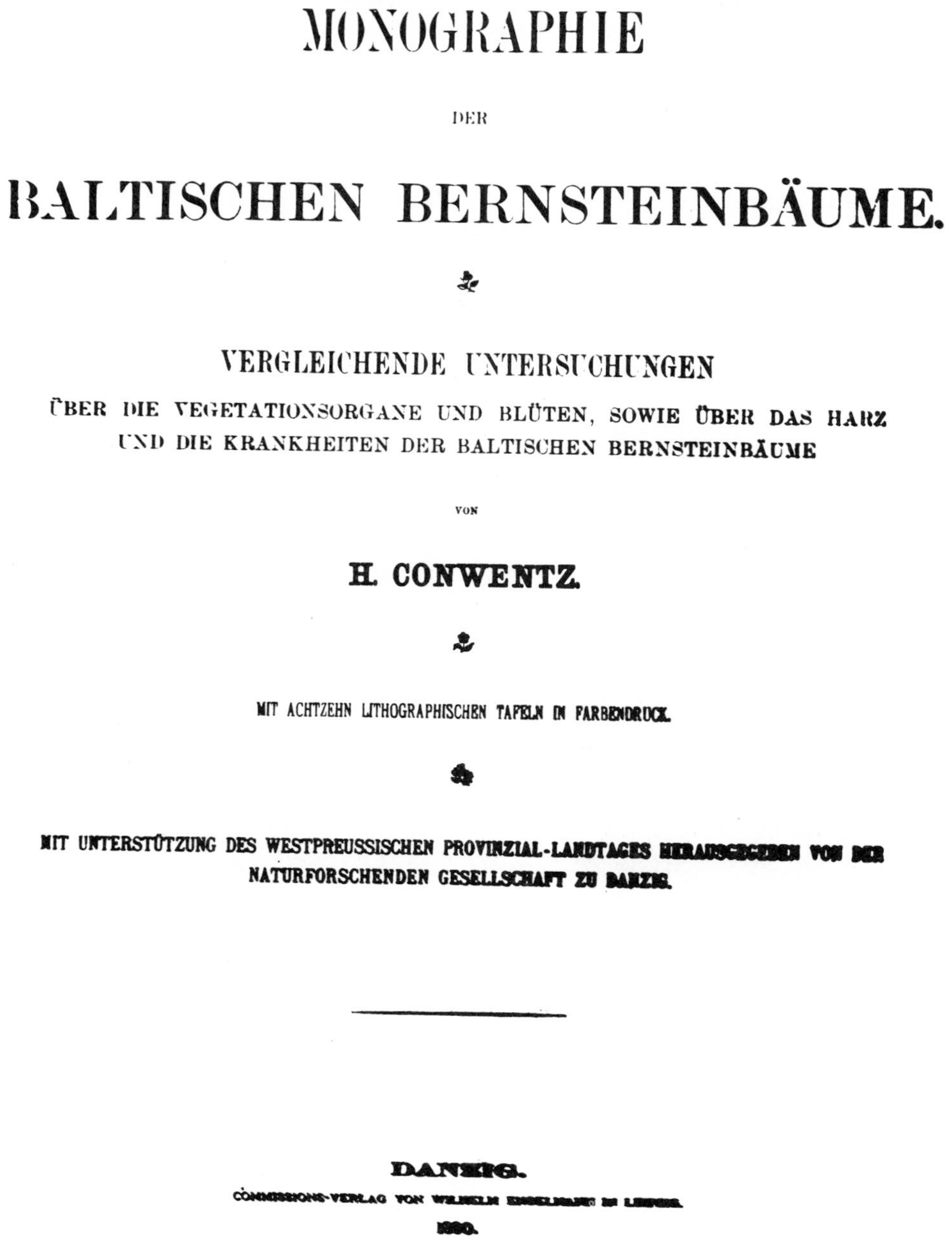

MONOGRAPHIE

DER

BALTISCHEN BERNSTEINBÄUME.

VERGLEICHENDE UNTERSUCHUNGEN

ÜBER DIE VEGETATIONSORGANE UND BLÜTEN, SOWIE ÜBER DAS HARZ
UND DIE KRANKHEITEN DER BALTISCHEN BERNSTEINBÄUME

VON

H. CONWENTZ.

MIT ACHTZEHN LITHOGRAPHISCHEN TAFELN IN FARBENDRUCK.

MIT UNTERSTÜTZUNG DES WESTPREUSSISCHEN PROVINZIAL-LANDTAGES HERAUSGEGEBEN VON DER
NATURFORSCHENDEN GESELLSCHAFT ZU DANZIG.

_____

DANZIG.
COMMISSIONS-VERLAG VON WILHELM ENGELMANN IN LEIPZIG.
1890.

Fig. 6-5. Title page from Conwentz's *Baltischen Bernsteinbäume* (1890).

the Lower Oligocene ended around 34 million years ago. This makes it possible that the amber-bearing forest existed over a span of 5 million years, as confirmed by finds of amber deposits within strata dated to these periods. During its span of existence, the boundaries of the forest changed, with the

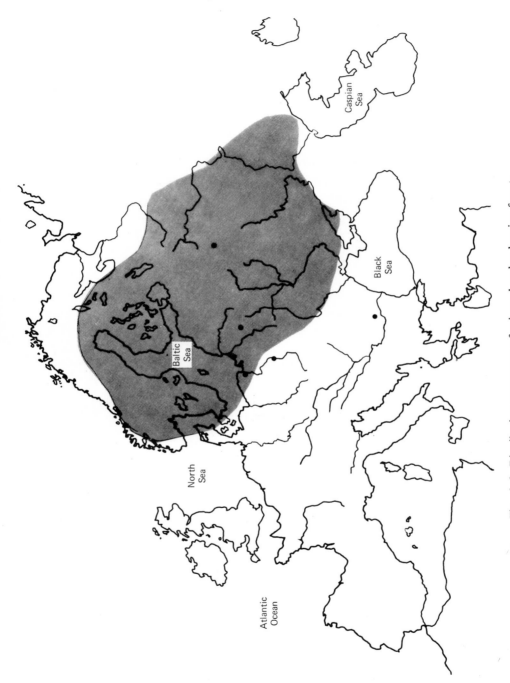

Fig. 6-6. Distribution-occurrence of primeval amber-bearing forests.

range both increasing and decreasing with the passage of time, its limits generally expanding toward the south. Northern Eurasia was subjected to repeated upheavals of the earth's crust during the forest's existence, with major changes in boundaries as prehistoric seas invaded the region, then regressed, causing continuous changes in the configuration of coastlines. At the same time, slow but continuous changes in climate may have altered the composition of the flora of the forests, yet trees continued to supply large quantities of amber resin. Fossil resins have been produced in various geological times and localities, but never in such vast quantities as found in the Baltic area, especially during the Oligocene epoch.

Knowing that amber was a product of forests and therefore of land origin, geologists were puzzled to find amber deposits occurring in marine sediments. Further geological studies provided reliable evidence that vertical movements of the earth's crust took place near the end of the Eocene and during the Oligocene in central and eastern European areas, which caused repeated marine transgressions and regressions. As a result, marine deposits were laid over land deposits, which, in turn, could have again been elevated to support a new forest and further deposits of amber. Kaunhoven,[13] in 1911, found evidence that the boundaries of marine and continental areas in Sambia during the Tertiary progressed through as many as 19 changes.[14]

The oldest Baltic amber deposits occurring in the lower blue earth layer were transported from the nearby land by the Eocene Sea. Streams flowing from amber-bearing forests carried the first deposits from the forest floor with the current as it emptied into the sea. As regions covered by amber-bearing forests were flooded during major marine transgressions at the close of the Eocene, soil containing amber deposits was eroded. Masses of amber, buoyant in agitated salt water, were carried by wave action and deposited on the sea bottom with sands and clay and the remains of sea fauna. Although amber occurs in a secondary bed, geologists from the Polish Academy of Sciences believe Eocene amber-bearing deposits are only slightly younger than the major amber lumps themselves. The secondary beds, however, are richer in amber than were the original beds of resin in the amber-bearing forest. Because sea waves continued to wash over the forest ground cover and soil, loosening and transporting lumps of resin to new marine beds, the secondary deposits continued to build up into thicker and greater amounts.[15]

The upper blue earth deposits of the Baltic were formed in a similar manner during the Lower Oligocene, but scientists are not certain if the primeval amber-bearing forests still existed or if they continued to produce resin during this epoch.

Successive sea transgressions washed out amber resin and transported it to new areas, and beginning with the Middle Oligocene, the invading sea attacked exposed fragments of amber from previously formed deposits and carried them to new places. For this reason, amber is now located in several strata in the Tertiary sequence from the Eocene to the Miocene. It is even found in brown coal deposits which were formed just prior to the Eocene and are associated with beds of lignite.

During the Ice Age, further transportation of amber took place, particularly during the Pleistocene epoch, when glaciers extended over the southern Baltic region near Sambia, cutting off considerable portions of the amber-bearing beds. Moraines, glacial rivers and sea currents carried the deposits over the European area and scattered them on sea bottoms and in land sediments of glacial origin. Therefore, glacial amber deposits appear in Quaternary sediments.

Today, amber continues to be deposited in new sites by the action of sea waves eroding sea bed amber deposits. Shore currents pick up loosened amber and toss it about in stormy periods, depositing it on the beaches.

In Samland (Sambia Peninsula), amber is found mainly in the blue earth layer deposited during the lower Oligocene, as dated by the occurrence of the index fossil oyster, *Ostrea ventilarum,* found associated with amber of the blue earth in Sambia. However, inland amber occurrences in Poland are in Quaternary sediments.

When new fossiliferous amber deposits are uncovered outside the Baltic region, geologists attempt to establish the age of the geological strata by studying the fossilized materials associated with the amber, as well as spore and pollen assemblages occurring within the amber. Details of the source and age have been established for some of the well-documented fossilized resin locations, such as the Simojovel Formation in Chiapas, Mexico (Oligocene-Miocene), the beach alluvium of Cedar Lake in Manitoba, Canada (reworked Cretaceous) and numerous sites in Cretaceous strata of Alaska. Amber varieties from other areas are assigned a geological age as indicated on the map shown in (Fig. 6–8). Recent research in the Soviet Union

Fig. 6-7. *Ostrea ventilarum,* index fossil of the Lower Oligocene found in the blue earth associated with amber in Sambia.

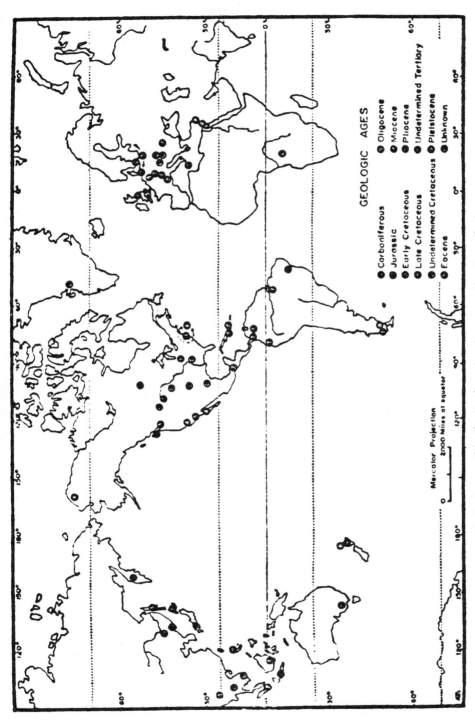

Fig. 2. Geographic distribution of ambers according to geologic age. © 1969 by the American Association for the Advancement of Science.

Fig. 6-8. Occurrences of amber and geologic age.

155

describes occurrences of amber and amber-like fossil resins in a number of locations within the USSR and gives an age range for these deposits from Early Cretaceous to Quaternary periods.[16]

CURRENT SURVEY OF AMBER DEPOSITS IN POLAND

Amber deposits occur in many regions of Poland, particularly along Baltic coastal areas. For centuries, amber has been collected on Baltic beaches, fished offshore and dug out of the ground. The largest quantities are collected from the sea shore or fished from the sea bed, and it is believed that such masses of amber are currently being washed out by the sea from the Tertiary deposits of Sambia and brought by currents to the Polish coast to again be deposited on the sea floor.

Underground amber deposits in Poland are almost exclusively found in Quaternary sediments laid down by glaciers, and occurrences coincide with the extreme range of Polish glaciation. Large quantities of post-glacial amber have been found in alluvial deposits inland in Pomerania, the Masurian Lake District and the Narew River Basin, but such deposits appear to be exhausted and amber is now seldom encountered farther south in Poland. On the other hand, Tertiary beds of amber, though to be an extension of the Samland deposits, occur in Pomerania at a depth of 10 or more meters and are found in Braniewo in East Pomerania, which is near Sambia, and in Mozdzanowo near Slupsk in Western Pomerania. The Polish Academy of Science reports that "attempts were made to exploit these deposits on a larger scale, but the costs incurred proved to be too high in relation to the obtained yield."[17] Compared to the rich amber beds of Sambia, deposits in Poland are low grade.

The following table prepared by the Polish Academy of Sciences in Warsaw (1974) shows the state of exploitation of amber deposits in Poland in the past and present (see Table 6–1).

REPORTS FROM THE USSR OF GEOLOGICAL SURVEYS OF THE AMBER DEPOSITS IN THE SOVIET UNION[18]

Amber deposits and amber-like fossil resins are found in numerous areas of the Soviet Union besides the Kaliningrad (Samland District of the USSR), which has produced amber since antiquity. As early as the 1700's, amber was discovered in the Northern European USSR, Siberia, the Ural Mountains, Kamchatka, the Carpathian Mountains and the Dnieper River Basin. The Carpathian and Dnieper basins produced a commercial quality amber (rumanite) used in jewelry for beads, amber buttons, and amber mouthpieces in the smoking industry. A 1929 survey of the Dnieper deposits along the right bank from the village of Starye Petrivtsy to Vyshgorod was conducted to assess the quantity of amber for possible use in jewelry and chemical industries. However, the amount of amber in the deposit did not

Table 6–1. Exploitation of amber deposits in Poland, past and present.*

Geological Age	Lithological Description of Bed	Important Exploitation Sites	Mode of Exploitation	Period of Exploitation	Present Condition of Bed	Remarks
Holocene	Muddy-silty delta deposits and sandy beaches.	Delta of the Vistula, the Vistula sandbar, Hel peninsula promontary.	Collection of beach amber and fishing offshore after stormy weather.	From the Neolithic to the present.	Still workable.	Yield to 4 tons/year. Mostly from the Vistula bar between Sobieszewo and Krynica Morska, "The Amber Shore."
Pleistocene	Gravels, sands and moraine clays, also river sands and gravels (alluvioglacial deposits).	Bursztynowa Gora near Gdańsk, Braniewo and its vicinity (Region of Sambia), the Narew Basin (Puszcza Kurpiowska Forest).	Primitive open-cut pits of the "box" type.	18th to 19th centuries.	Exhausted.	At Bursztynowa Gora, signs of mining date to some hundreds of years ago. In the Puszcza Kurpiowska Forest over 100 small pits with a good output were worked during the 19th century.
Pliocene	Sandy-silty deposits of the brown coal sediments.	Bory Tucholskie Forest.	Open cuts in shallow depressions where brown coal had been mined.	1835 to 1865.	Exhausted.	"Nests" of amber were of such output that over 200 workmen used to be at work at one time.
Miocene	Deposits of brown coal deposits: fine-grained quartz sands, mostly with a high admixture of mica, also sandy silts.	Region of Slupsk (Mozdzanowo, Starkowo, Ugoszcz), Bretowo near Gdańsk-Wrzeszca, Braniewo (region of Sambia), Masurian Lake District, Narew Basin (Puszcza Kurpiowska Forest).	In Pomerania, small pits varying in depth with shafts provided with brattice; in the Masurian Lake District and the Narew basin, small open pits of the "box" type.	18th to 19th centuries.	Exhausted.	In the Slupsk region, the output of amber was abundant. Over 100 workmen used to be at work at one time. At Bretowo there were three pits worked until 1940.
Lower Oligocene	Glauconitic sands of the amber formation (upper "blue earth").	Mozdzanowo near Slupsk (Western Pomerania).	Deep boring (the amber-bearing beds occur at a depth between 100 to 105m).	1950 to 1956.	Abandoned.	Deposits rich but costs of exploitation too high.
Tertiary(?)	Lack of documentation.	Offshore belt of the Bay of Gdańsk, under water and on shore, Stogi near Gdańsk.	Below the bottom of the Bay of Gdańsk by dredging; on shore by extractor pumps.	1970 to 1972.	Exploitation not continuous.	The accumulated mass of amber may be a fragment of Tertiary deposits, or an extension to the west of the Sambian beds

* Polish Academy of Science, Warsaw 1974, p. 80.

exceed 7 grams per cubic meter and it was not considered economical to mine.

Earlier geological field trips in the middle of the nineteenth century discovered amber-like fossil resins in the eastern and southern regions of the Soviet Union. Reports describe grain size, color and transparency, but not the quantity of amber contained in the deposits. Amber of the eastern Caspian Sea area is a pale champagne-yellow and transparent. (See Color Plate 35 for Caspian Sea piece.) It is not generally used commercially because of its softness and brittleness. Amber has also been found along the shore of the Arctic Ocean, and these ambers were used as a substitute for labdanum (ladanum), a dark colored brittle resinous substance from Old World plants known as rock rose (genus, *Cistus*), and are also used in flavorings and perfumes.

In 1936 and 1937, extensive work was begun in Transcaucasia for treatment and extraction of fossil resins found near the village of Shasha, but the small amounts found made the work unprofitable. Also in this period, amber was found in the Soviet Far East while processing coal, with the amber separated by hand during the treatment of the coal. Only 8 to 10 percent of the total material mined was amber and it was not considered useful for the jewelry industry. However, a fossil resin found on the eastern coast of southern Sakhalin and on the beach near the villages of Vzmor'ye, Firsovo and Stravodubskoye is both distinctive and unusually decorative. Although these resins are not succinite, they are of some interest to the jewelry industry.

Scientific studies of the amber-like fossil resins found within the USSR show most are not succinite; in fact, in each region, different types of fossil resins may be found together. For example, in the Kaliningrad District, six types of fossil resins occur: succinite, gedanite, glessite, stantienite, beckerite and krantzite. The succinite, or the classical Baltic amber, makes up 90 percent of the total yield. In the Carpathian area, rumanite, schraufite and delatynite are found along with succinite.

After World War II, Soviet geoloists from 1948 to 1957 studied amber reserves in the Kaliningrad District of the Baltic, and though it was reported that the reserves in the deposit were defined, figures were not released. In 1964, succinite was found in the Byelorussian SSR and Ukranian SSR which proved to be suitable for working into jewelry.

Soviet authorities suggest a further increase in amber reserves is possible not only by continued mining on land, but also through exploration of the shelf of the Baltic Sea near the western and northern coasts of the Kaliningrad District (Samland), especially along the western coast south of Yantarnyy (Palmnicken) where 75 percent of all amber extracted in the Baltic area has been produced. [19,20,21,22]

The 1937 publication of *Bernstein* by Karl Andrée[23] graphically illustrates the process of extraction of amber from the earth, the manufacturing processes through which amber progressed, and its various uses. (A translation of this amber flowchart is found in Table 6–2.) Currently, amber sources are not as plentiful, and production is not on the same large scale.

INCLUSIONS IN AMBER

Systematic investigation of plant and insect inclusions in amber began in 1830, when the German naturalist Georg Berendt[24] examined over 2000 specimens. A. Menge and H. R. Goeppert[25] joined him in 1845. As a result of their combined work, Goeppert reported to the Berlin Academy in 1853 163 different species of vegetable remains, divided into 24 families and 64 genera.

Table 6-2. Amber flow chart: An overview of the extraction and uses of amber.*

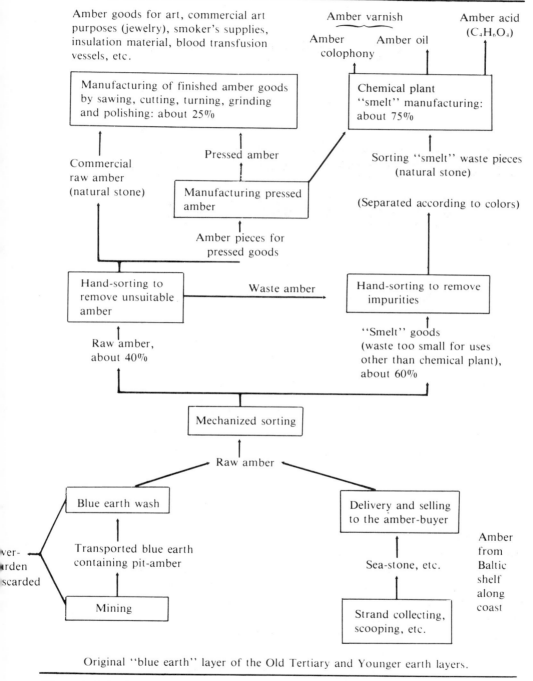

Amber goods for art, commercial art purposes (jewelry), smoker's supplies, insulation material, blood transfusion vessels, etc.

Amber varnish

Amber colophony   Amber oil

Amber acid (C₄H₆O₄)

Manufacturing of finished amber goods by sawing, cutting, turning, grinding and polishing: about 25%

Chemical plant "smelt" manufacturing: about 75%

Commercial raw amber (natural stone)

Pressed amber

Manufacturing pressed amber

Sorting "smelt" waste pieces (natural stone)

(Separated according to colors)

Amber pieces for pressed goods

Hand-sorting to remove unsuitable amber

Waste amber

Hand-sorting to remove impurities

Raw amber, about 40%

"Smelt" goods (waste too small for uses other than chemical plant), about 60%

Mechanized sorting

Raw amber

Blue earth wash

Delivery and selling to the amber-buyer

Amber from Baltic shelf along coast

Transported blue earth containing pit-amber

Sea-stone, etc.

ver-rden scarded

Mining

Strand collecting, scooping, etc.

Original "blue earth" layer of the Old Tertiary and Younger earth layers.

*From Andrée, *Bernstein*, 1937.

Die

im Bernstein

befindlichen

ORGANISCHEN RESTE DER VORWELT

gesammelt

in Verbindung mit Mehreren bearbeitet

und

herausgegeben

von

Dr. Georg Carl Berendt.

praktischem Arzte zu Danzig.

Director der naturforschenden Gesellschaft zu Danzig, Ehrenmitgliede des böhmischen Museums zu Prag und correspond. Mitgliede der schlesischen Gesellschaft für vaterländische Cultur zu Breslau, der Gesellschaft für Natur- und Heilkunde zu Dresden, der naturforschenden Gesellschaft zu Görlitz, der physikalischen Gesellschaft zu Königsberg, der entomological Society zu London, der Kaiserl. mineralogischen Gesellschaft zu St. Petersburg, des wissenschaftlichen Vereines zu Posen und des entomologischen Vereines zu Stettin

Fig. 6–9. Title page of Berendt's *Bernstein* (1845).

In 1890, H. Conwentz[26] published a large volume on the results of his studies on inclusions in amber. Being the first to examine the rare pine enclosures such as needles, fragments of wood and pollen grains, he concluded that at least four different species contributed to the formation of the Baltic amber. Because so few fragments of pine are preserved in amber, and distinction of the particles now detached from parent trees is so difficult, it is almost impossible to identify exact species. In view of these difficulties, Conwentz recommended that all prehistoric conifers which produced the amber resin be described with one collective name, *Pinus succinifera,* or "amber-bearing pine," and this name was adopted by other scientists and is still in use.

Runge[27] found 174 different species of flies, ants, beetles and moths; 73 species of spiders; and many species of centipedes in amber. Such imbedded

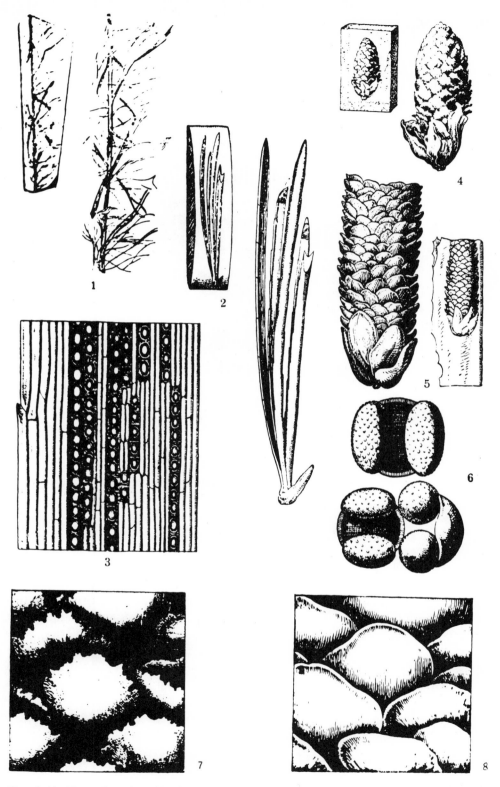

Fig. 6–10. Examples of conifer inclusions in Baltic amber *(Pinus succinifera)*. (Plate from Conwentz, *Monograph on Baltic Amber,* 1890.)

Fig. 6-11. Inclusions in Baltic amber of botanical particles from prehistoric conifers which grew during the Eocene and Oligocene (actual size: 5mm). (Author's collection.)

specimens of extinct life are often found with details such as compound eyes, wing veins, scales on insects and hairs on the legs of spiders clearly visible. It was speculated that insects were attracted to the sticky, sweet surface of the resin as it flowed over the bark in its liquid state. When these insects became engulfed in the viscous material, their bodies were preserved just as they were without serious damage. In some specimens, only fragments of insects, such as antennae, wings or legs, are found, which suggests they were torn off as the insects attempted to extricate themselves from the sticky mass. It is noteworthy that insects are most often found in shelly amber, a kind which formed from numerous layers of resin, corresponding to successive flows of resin.

Microscopic examination of insects preserved in amber shows hollow body cavities where internal organs have decayed. What is actually observed is a mold in the amber, lined with a pigment composed of metamorphosed and carbonized material from the tough horny outer covering (exoskeleton) of the insect. It is for this reason study of such insects must be done in-site; if the supporting amber were dissolved away, the insect would crumble into dust-like fragments.[28]

Fig. 6-12. Wood fibers from prehistoric trees growing approximately 40 million years ago and now enclosed in Baltic amber (actual size: 12mm). (Author's collection.)

Insect inclusions in Baltic amber indicate that the climate 40 to 60 million years ago was much warmer than at the present time, and many of the species found no longer inhabit the cool Baltic region, but resemble those now found only in the temperate climes of Europe and North America. Very few of these species exist in their exact form today, though most survive in modern variants.

Of special interest to scientists are the ants found in Baltic amber because they give much information about the evolution of insects. Not only were they more abundant 40 million years ago than they are now, but some types are now extinct or are no longer found in the Baltic area. One species, now found only in Ceylon, weaves leaves together with fine threads to form its nest. Having no silk-producing organs, the ants hold their larvae between their mandibles, gently squeezing the larvae to encourage spinning. Other ants hold the leaves in their mandibles until the threads dry and the leaves are secured in place. These curious ants occur in Baltic amber and one was even found still holding a larva.[29] The ant most abundantly found in Baltic amber, however, is similar to the most common ants in Europe and North America today, the mound-building black ant, *Formica fusca*.[30]

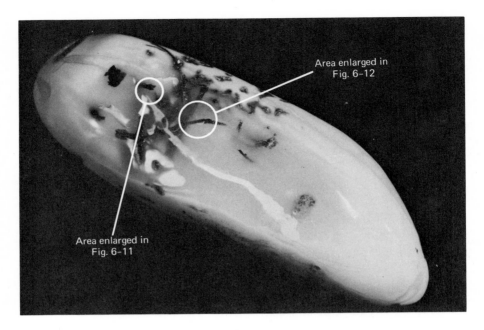

Area enlarged in
Fig. 6-12

Area enlarged in
Fig. 6-11

Fig. 6/12A Area enlarged in Fig. 6–11.   Area enlarged in Fig. 6–12.

Though flies make up to 54 percent of all the insects trapped in amber, they today account for a considerably larger portion of the total insect population than they did 40 million years ago. Most examples found entombed in amber are of small varieties, possibly because it was more difficult for these to pull themselves free from the sticky mass than it was for larger insects.[31] Spiders and termites are also found in amber. Again, these are indicative of a warmer climate, as termites are no longer found in Northern Europe.

In 1911, Dr. Richard Klebs,[32] after collecting insect inclusion specimens of amber for 40 years, surveyed the extensive Stantien and Becker collection at the Königsberg University Geological Institute Museum and estimated the proportions of enclosed insects as follows:[33]

Diptera (two-wing flies)	50.0%
Pseudoneuroptera (termites, mayflies)	10.7%
Rhynchota (lice, gnats)	7.1%
Neuroptera (caddis)	5.6%
Hymenoptera (bees, wasps, ants)	5.1%
Arachnoidea (spiders, mites, scorpions)	4.5%
Coleoptera (beetles)	4.5%
Orthoptera (cockroach, grasshopper)	0.5%
Microlepidoptera (little moths)	0.1%
Various	1.1%

Der Bernstein in Ostpreußen.

~~~~~~~~~

Zwei Vorträge

von

Wilhelm Runge.

Mit einem Titelbild und 10 in den Text eingedruckten Holzschnitten.

Berlin, 1868.

C. G. Lüderitz'sche Verlagsbuchhandlung.
A. Charisius.

Fig. 6-13. Title page of Runge's *Der Bernstein in Östpreussen* (1868).

The collection studied by Klebs included about 120,000 of the finest amber specimens selected during amber mining and contained inclusions of flora and fauna. At least 70,000 arthropod inclusions were found among these specimens. It was thought that this entire collection was destroyed by fire during World War II,[34] but, fortunately, reliable sources report this is not

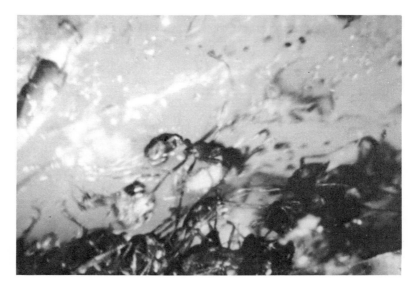

Fig. 6-14. Primitive ant inclusions in amber from the Caspian Sea region. As many as 36 species of ants from the family *Formicidae,* belonging generally to tropical climate forms, have been described in Baltic amber. (Microphotograph.)

true. In 1965, the German scientist, Hennig, in his study of Diptera in Baltic amber, indicated the collection was still available for study. According to him, it was moved during the war for safe-keeping to the Geological-Paleontological Institute of the University of Göttingen, Germany, and is still there. That portion of the University of Königsberg collection assembled by Berendt is now housed in the Paleontological Museum of Humboldt University, Berlin, while some specimens are reported to be in the British

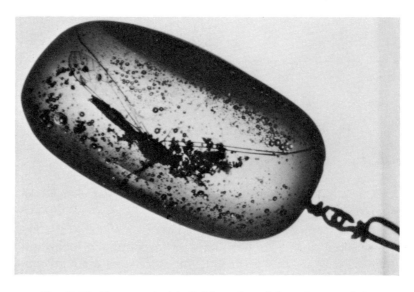

Fig. 6-15. Fly entombed in Baltic amber. (Microphotograph.)

Fig. 6–16. Spider entombed in Baltic amber. (Microphotograph.)

Museum (Natural History) Department of Paleontology. Thus, the collection appears not to be lost forever, but is available for future generations.[35]

Insect finds in Baltic amber occur in about one piece per thousand. However, fauna inclusions are more abundant in amber found in the Dominican Republic from the Lower Miocene formation approximately 30 million years ago, with about one piece in a hundred containing insects. Ants and flies appear most frequently in both Baltic and Dominican types. Insects included in Dominican amber are generally larger than those found in Baltic amber, but are rarely larger than 2 inches. One Dominican amber source reports having discovered a large butterfly, measuring 5 inches across the wing spread, perfectly preserved in amber, but this is most unusual and appears to be the largest on record having been found.[36]

Flora and fauna inclusions in amber are used to piece together a picture of the way amber forests may have looked. It is impossible to know exactly, since amber occurs in secondary deposits, and not all forms of life in existence during the resin-producing period were trapped in amber.

About 200 species of spore-producing (gymnosperms) and seed-producing flowering plants (angiosperms) have been identified from remains enclosed in amber. This suggests the vegetation was both rich and varied. Identified gymnosperm enclosures in amber include firs, cypresses, junipers, pine, spruce and Arbor vitae. Angiosperms are represented by numerous species of oaks, beech, maple, chestnut, magnolia and cinnamon. Remains of palms with fan-shaped as well as pinnate leaves have been identified, too, and numerous ferns, mosses and flowering herbacious plants apparently also formed a ground cover in the ancient forests. Having knowledge of the present habitats and natural environments of various species of plant and animals similar to those found in the amber, scientists are able to describe

Der Bernstein.

Seine Gewinnung, Geschichte und geologische Bedeutung.

Erläuterung und Catalog der Bernstein-Sammlung
der Firma

Stantien & Becker.

von

Richard Klebs.

Königsberg i. Pr.

Druck von S. Landien in Königsberg i. Pr.

Fig. 6-17. Title page of Klebs.

the environment of the amber-producing forests. Fungi inclusions suggest that humid areas comprised portions of the forest, though ants and other insects indicate there were also dry regions. The presence of beetles indicates some amber-bearing trees grew on high mountainsides, as these insects are characteristically found in areas of swift-running water. Stagnant ponds were also present in some locations, as evidenced by gnats and their larvae. Fleas and gadflies indirectly prove this was the era of the early mammals,

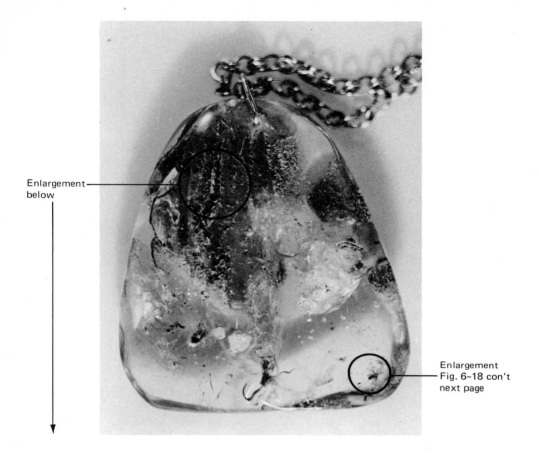

Enlargement
below

Enlargement
Fig. 6-18 con't
next page

Fig. 6-18. Pendant of Baltic amber with seven insect inclusions, a centipede and insect fragments. Enlargement of the centipede *(Chilopoda)*; actual size: 9mm.

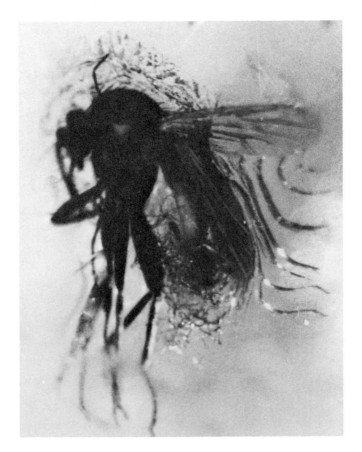

Fig. 6–18 (*continued*)
Enlargement of a mosquito or midge inclusion (superfamily *Tipuloidae*); actual size: 2mm.

since these insects are parasites on mammals. And the presence of wood-boring insects indicate particular kinds of trees that were present in the amber forest.[37]

Beetles from the family *Anobiidae* are often found in Baltic amber. About 40 different species have been described, with the most common belonging to the genus *Anobius*. Today, there are about 1,200 species of *Anobiidae*, most living in dry wood, bark of trees and cones of conifers, and in tree pith in tropical zones. Most live in communities and lead a parasitic life. The amber from the Caspian sea region shown in Fig. 6–20 encloses a flat-headed wood-boring beetle from the family *Platypodidae*, or pin-hole borers. Such beetles are very slender, with elongated parallel sides, and prefer deciduous trees. Several beetles found in the illustrated piece were associated with a cluster of primitive ants.

As previously mentioned, termites and other insects, along with palms and tropical shrubs, suggest that the climate was subtropical to tropical. On the other hand, numerous plant inclusions characteristic of the temperate climate zones are found, too, such as the pine, larch and maple, suggesting

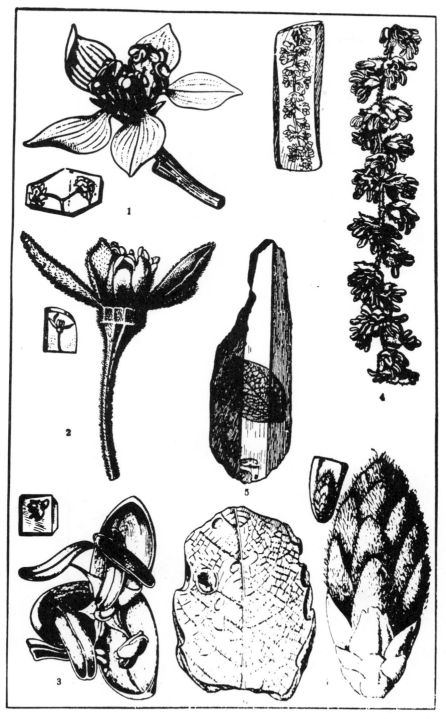

Fig. 6–19. Example of angiosperm remains from Baltic amber. (Plate from Conwentz mono-graph of 1890.)

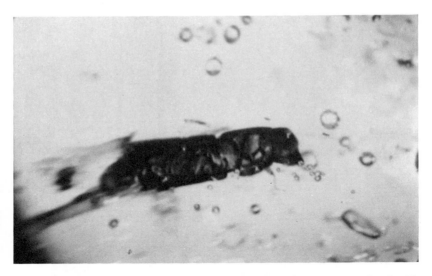

Fig. 6-20. Wood-boring beetle in Caspian Sea amber (order *Coleoptera,* family *Platypodidae.*) (Microphotograph.)

that the forest also may have covered a mountainous region where the climate was cooler as a result of the elevation.

Some lumps of fossil resins also bear surface impressions of leaves, preserving details of vein and cell structure. In other specimens, surface textures

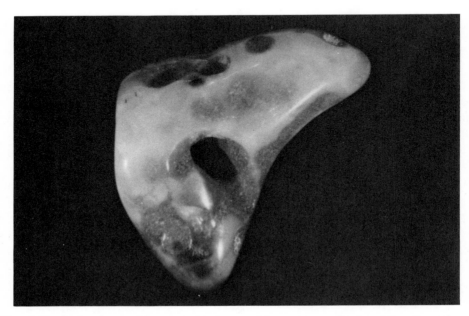

Fig. 6-21. Amber lump with a natural hole caused by enclosure of twigs at time the resin was exuded. As time passed, the twig decayed and crumbled away. About 3 inches by 4½ inches. *(Courtesy Amber Forever, Florida.)*

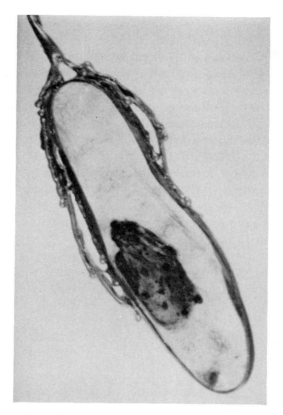

Fig. 6-22. Baltic amber pendant with a shell enclosure. *(Courtesy Amber Forever, Florida.)*

provide clues to the environment in which the amber came to rest. For example, lumps found in sandy, dry, well-aerated areas, such as just below the surface of a dune, tend to be thick-crusted. Since sand provides such lumps with free access to the atmosphere, oxidation occurs, resulting in the formation of a crust or "bark," the latter deepening in color with age and gradually thickening to the point where it readily flakes off. Amber masses from diluvial deposits laid down during the Pleistocene may show striations or cracks formed as the pieces were transported by glacial debris. Amber lumps with natural holes in them are especially interesting, apparently having been formed by amber resin dripping onto, and surrounding, fragments of twigs. Such twigs became partially enclosed by resin and in time decayed and crumbled away, leaving tubular voids in the amber.

Another surface feature of amber which rested for some time on the sea floor is a covering of barnacles or other skeletons of colonial crustaceans. Sedentary barnacles, *Balanus improvisus,* living in the offshore zone of the sea floor at that time, settled on the amber lumps and became so firmly attached that they must be ground off before cutting and polishing.

MODERN BOTANICAL RESEARCH RELATED TO AMBER

Up to the mid-1960's, scientific studies of amber mainly were confined to identification and classification of insect and plant inclusions. However, the discovery of new analytical methods, along with new approaches to the study of amber, provided powerful tools for investigation of the substance of amber rather than its inclusions. Infrared (IR) spectrophotometry, X-ray diffraction, scanning electron microscope photography, mass spectroscopy, and gas liquid chromatography came into use to provide more information on the structure and composition of amber. Botanists particularly became interested in the chemical composition because resins are mixtures of complex terpenoid compounds produced by plants, with the proportions varying from one plant species to another. Studies comparing the compositions of amber to recent resins produced by present day plants delineated possible evolutions of amber-producing plants.

Analysis of amber by IR spectrophotometry is used extensively to compare such constituents to those found in recent resins. Spectra thus obtained are examined to see if absorption peaks in amber can be matched to modern resins, and thus suggest a common botanical origin. Jean Langenheim,[38] of the University of California, worked extensively in this area of investigation and matched a large number of ambers from various deposits to present day tree resins. In 1969, she published the botanical affinities (see Table 6–3).

Table 6–3. Age, geographic location, geological occurrence and botanical affinities of a large sample of recorded deposits of amber.[*]

Location	Geologic Occurrence	Botanical Affinities
Carboniferous		
Northumberland, England		Coniferales
Upper Mississippi Valley		Coniferales
Montana		Coniferales
Jurassic		
Bornholm		Coniferales
Lower Cretaceous (Albian Stage)		
Maryland (Upper Patapsco Formation)		Araucariaceae, Taxodiaceae-Cupressaceae, Pinaceae
Upper Cretaceous (Cenomanian-Turonian; Turonian-Coniacian)		
Magothy River, Maryland (Raritan and Magothy Formations) Martha's Vineyard, Massachusetts (Raritan, Magothy Formations)		Araucariaceae, Taxodiaceae

Table 6-3 *(continued)*

Location	Geologic Occurrence	Botanical Affinities
	Upper Cretaceous *(Cenomanian-Turonian)*	
Cliffwood, Bordentown, New Jersey (Raritan Formation)		Araucariaceae
Kreischerville, New York (Raritan Formation)		Taxodiaceae, Pinaceae
	Upper Cretaceous	
St. Georges, Delaware		
Kincona, New Jersey		
Roebling, New Jersey		
Harrisonville and Pemberton, New Jersey		Hamamelidaceae (*Liquidambar*)
Cedar Lake, Manitoba		Araucariaceae
	Upper Cretaceous (Turonian)	
Kuk River Drainage, Alaska		Taxodiaceae
	Upper Cretaceous (Campanian)	
Baja California, Mexico		
	Cretaceous	
Hardin County, Tennessee		
Black Hills, South Dakota		
Cañon Diablo, Arizona		
Terlingua Creek, Texas		
Ellsworth County, Kansas (Kiowa Formation)		
Peace River, British Columbia		
Hare Island, Greenland		
Vienna, Austria		
Hungary		
Switzerland		
Bahia, Brazil (Reconcavo group)		
	Paleocene-Eocene	
S.E. Coast, England (London Clay formation)		Burseraceae
	Eocene	
Hukawng Valley, Bruma		
	Eo-oligocene	
Seattle, Washington (Renton Formation)		
	Eo-oligocene; Miocene	
Baltic Coast		Pinaceae (*Pinus*)
	Eocene (Domengine)	
Simi Valley, California		Pinaceae

Table 6-3 (*continued*)

Location	Geologic Occurrence	Botanical Affinities
	Eocene-Miocene	
Rhineland Brown Coals		Hamamelidaceae (*Liquidambar*) Taxodiaceae
	Oligocene	
Argentina (Patagónica Formation) Dominican Republic Savoy		
	Oligo-Miocene	
Chiapas, Mexico (Simojovel Formation)		Leguminosae (*Hymenaea*)
	Miocene	
Pará, Brazil (Pirabas Formation) Carpathian Mountains, Rumania Central Sumatra		Leguminosae (*Hymenaea*) Dipterocarpaceae
	Miocene; Pliocene	
Auckland Province, New Zealand		Araucariaceae (*Agathis*)
	Pliocene	
Victoria, Australia Java Luzon, Philippine Islands		Araucariaceae (*Agathis*)
	Tertiary	
Medellín and Girón, Columbia Magdalena, Chile Guayaquil, Ecuador Northern Sicily S.E. Borneo Hokkaido, Inotani, Japan Sakhalin Haiti		Leguminosae (*Hymenaea*) Burseraceae (*Protium*)
	Pleistocene	
Tel-Aviv, Israel		Anacardiaceae (*Pistarca*)
	Pleistocene-Recent	
N.E. Angola, Africa		Leguminosae (*Copaifera*)
	Unknown Age	
Vladivostok Siam Cochin-China Manchuria Kamchatka		

X-ray diffraction is used to study organic complexes such as amber in order to detect and identify crystalline components. By such means, Judith Frondel,[39] of Harvard University, detected a substance called alpha-amyrin in several amber samples, including Baltic amber. Alpha-amyrin is a major component in resins from some species of angiosperms (flowering trees). Since Baltic amber was considered exclusively a product of coniferous evergreens, or gymnosperms, this new data cast an entirely new light on the amber source of the primeval amber forest. However, composition has not yet been determined accurately enough to do more than indicate the botanical affinities with present day trees. At the present time, botanists are not sure why plants produce resins, some believing them to be secondary products of normal metabolism, others suggesting that resins are produced as a result of injury or disease; still others believe that resins are exuded to attract or repel insects.[40]

Amber in Archaeological Finds

The composition of amber not only yields useful information to botanists but is valuable to archaeologists in determining the origin of amber artifacts. Amber containing succinic acid is considered to be of Baltic origin, whereas amber without succinic acid (retinite) is determined not to be from the Baltic source. It was Otto Helm[41] who believed Baltic amber was uniquely distinguished by its succinic acid and thus a historical milestone in archaeology. However, there are limits to its usefulness. Though Helm was correct in claiming all Baltic amber contains from 3 to 8 percent succinic acid, this does not mean ambers other than the Baltic contain less or no succinic acid. In fact, ambers from other localities may contain comparable amounts. On the other hand, the test for succinic acid did rule out a Baltic source if no succinic acid was found in an unknown specimen. Another drawback to testing for succinic acid is that it requires destruction of more than a gram of amber artifact because of the chemical analytical method used, requiring heating the amber until it decomposes.

C.W. Beck summarizes how modern laboratory methods can be useful in amber studies, stating:

The success of modern instrumental analysis to decide questions of archaeological provenance by means which are either nondestructive or which use very small samples has opened new approaches to the amber problem. Among the options, infrared spectroscopy has proven both decisive and convenient. Infrared spectra do not distinguish unequivocally among all of the many non-Baltic European resins, probably because many of them have the same botanical sources. But they do permit the positive identification of Baltic amber and thus allow the direct recognition of imports from the north among the large number of amber finds in European, and particularly in Mediterranean, archaeology.[42]

Because archaeological samples are often contaminated with other substances, IR peaks are sometimes obscured, making identification impossible. An archaeological study group directed by Beck at Vassar College developed a computer technique to analyze IR spectra to aid in identification. This technique was successfully applied to amber artifacts collected from various Greek sites, including Mycenae, and provided evidence that as early as 1550 B.C., Baltic amber originating in the northern regions of Europe was used in Greece in the Mediterranian region.[43]

Beck's experiments tested other analytical methods which could aid in determining the provenance of fossil resins.[44] Besides IR spectroscopy, pyrolysis (destructive chemical breakdown by heat), followed by gas chromatography and analysis for trace elements, proved useful. Specimens with too much oxidation on the surface do not produce characteristic spectra as a result of the chemical changes which have taken place. For this reason, other methods are necessary for determining the type of resin.

REFERENCES

1. Ley, Willy, *Dragons in Amber*. New York: The Viking Press, 1951, p. 20.
2. *Ibid.*, p. 19.
3. Sendelio, Nathanaele (Nathanael Sendel), *Historia Succinorum*. Elbing, 1742.
4. Lomonosov, Mikhail V., [*Complete Collection of Works* 5]. Moscow-Leningrad, 1954, p. 389.
5. Bock, Friedrich S., *Versuch einer Kurzen Naturgeschichte des preussischen Bernstein.* Königsberg, 1767.
6. Heer, Oswald, *Die Tertiäre Flora der Schweiz,* Bd. 3., Winterthur, 1859; *Miocän baltische Flora,* Königsberg, 1896.
7. Zaddach, G., *Das Tertiärgebirge Samlands.* Königsberg: Phys. Ökon. Ges., 1867, pp. 85–197.
8. Zalewska, Zofia, *Amber in Poland, Guide to the Exhibition.* Warsaw: Polish Academy of Sciences, 1974, p. 57.
9. Berendt, G., *Der im Bernstein befindlichen Organischen Reste Der Vorwelt.* Berlin, 1845.
10. Heer, *Die Tertiäre Flora.*
11. Conwentz, H., *Monographie der baltischen Bernsteinbäume,* S. 151, Taf. 18. Danzig, 1890.
12. Komarow, W. L., *Proiskhozhdenie Rastenii,* Led. 7. Moscow, 1943.
13. Kaunhoven, F., *Der Bernstein in Östpreussen.* Berlin: Jahrb. K. Preuss Geol. Landesanstalt Bd. 34. T 2, 1913.
14. Zalewska, *Amber in Poland,* p. 74.
15. *Ibid.*, p. 75.
16. *Ibid.*, p. 80.
17. *Ibid.*
18. Yushkin, N. R. and Sergeyeva, N. Ye., "Textures of Amber from Yugorskiy Peninsula," *Doklady Akademii Nauk S.S.S.R., Earth Science Section* **216**, *1–6:* 151–153, Washington, D.C. 1974.
19. Savkevich, S. S., "State of Investigation and Prospects for Amber in U.S.S.R., "*International Geology Review* **17**, *8:* 920, Washington, D.C., 1975.
20. Savkevich, S. S. and Popkova, T. N., "New Data on Amber From the Right-bank Area of the Kheta and the Khataga Rivers," *Doklady Akademii Nauk S.S.S.R., Earth Science Section* (Minerology) **208**: *1–6:* 131–132, Washington, D.C., 1973.
21. Srebrodol'skiy, B. I., "Amber in Sulfur Deposits," *Doklady Akademii Nauk S.S.S.R., Earth Science Section* **223**, *1–6:* 204–205, 1975.

22. Velikiy, N. M., "Amber finds on the Northwest Shore of the Aral Sea," *Doklady Akademii Nauk S.S.S.R., Earth Science Section* **221**, *1-6:* 164–166, Washington, D.C., 1975.

23. Andrée. Karl, *Der Bernstein und seine Bedeutung in Natur und Geisteswissen-schaften, Kunst und Kunstgewerbe.* Technik, Industrie und Handel. Königsberg, 1937.

24. Berendt, *Organische Reste im Bernstein.*

25. Goeppert, H. R. and Menge, A., *Die Flora des Bernsteins und ihre Beziehungen zur Flora der Tertiärformation und der Gegenwart,* Bd. 1, Danzig, 1883, 63 pp.

26. Conwentz, *Baltischen Bernsteinbäume.*

27. Runge, W., *Der Bernstein in Östpreussen.* Berlin, 1868.

28. Brues, C. T., "Insects in Amber," *Scientific American* **185**, *5:* 57, 1951.

29. Ley, *Dragons in Amber,* pp. 45–46.

30. Brues, "Insects in Amber," Scientific American **185**, *5:* 57.

31. *Ibid.,* **185**, *5:* 58.

32. Klebs, R., *Der Bernstein.* Königsberg, 1880.

33. Klebs, R., *Der Bernsteinschmuck der Steinzeit.* Königsberg, 1882.

34. Ley, 1951; Pinkus and Gudynas, 1963; *Encyclopedia Lituanica,* 1974.

35. MacAlpine, J. F. and Martin, J. E. H., "Canadian Amber—A Paleontological Treasure-Chest," *The Canadian Entomologist* **101**: 819, August 1969.

36. Zahl, Paul, "Amber, Golden Window on the Past," *National Geographic* **152**, *3:* 431, September 1977.

37. Zalewska, *Amber in Poland,* p. 67.

38. Langenheim, Jean H., "Amber: A Botanical Inquiry," *Science* **163**, *3:* 1156–1167, 1969.

39. Frondel, Judith W., "Amber Facts and Fancies," *Economic Botany* **22**, *4:* 381, October—December 1968.

40. "Amber—Substance of the Sun," *Chemistry* **45**: 21–22, April 1972.

41. Helm, Otto, *Schriften Naturforschenden Gesellschaft.* Danzig: N. R., Nr. 2, 1885, pp. 234–239.

42. Beck, Curt W., "Archaeological Chemistry," *Science and Archaeology* (Brill, R. E., ed.). Cambridge: M.I.T. Press, 1971, p. 235.

43. *Ibid.,* pp. 235–240.

44. Beck, Curt W., "The Origin of the Amber found at Gough's Cave, Cheddar, Somerset," *Proceedings, University of Bristol Spelaecological Society* **10**, *3:* 272–276, 1965.

7

Amber Varieties—Fossil Resins Other Than Baltic Succinite

Amber from the Dominican Republic

The Dominican Republic has been a known source for amber in North America longer than any other amber deposit in the Western Hemisphere. In Columbus' account, in 1496, of his Second Voyage to the New World, he states that amber was found on the island, then known as Hispaniola, and describes a mining region near the Tower of Conception, a fortress located on the border of the country ruled by Guanahanis (Guarionexius).

Though this amber was mentioned in the literature in the nineteenth century, it was not exported commercially. In 1891, J. G. Haddow [1] noted:

At Santiago, in San Domingo, in the valley of the brook Acagua, amber pieces, some as large as the egg of a goose, reward the explorer. The Acagua brook carries away the amber from hills of marl which is rich in petrifactions, and bears a near resemblance to miocene clay of the Vienna basin.

In 1905, amber containing inclusions of decayed twigs was described as originating from an amber-bearing formation in the Monte Cristi Range*,[2] but despite early knowledge of the source, Dominican Republic amber began to be exported on a commercial scale only recently.

The main Dominican deposits are located in two concentrated areas in the Monte Cristi Range along the northern border of the Vega Real central plains area, north of Santiago between Altamira and Canca (see Fig. 7-1). The largest and original mining site is northwest of Tamboril in the Peña region. The location is between two gorges, Los Meninos and Perez, of the Arroyo Capancho tributary of the Rio Gurabo. The amber is found at an approximate elevation of 1240 meters, and mountainous terrain makes mining difficult.[3] The second site is below Pico Diego de Ocampo, near

---

* Cordillera Septentrional

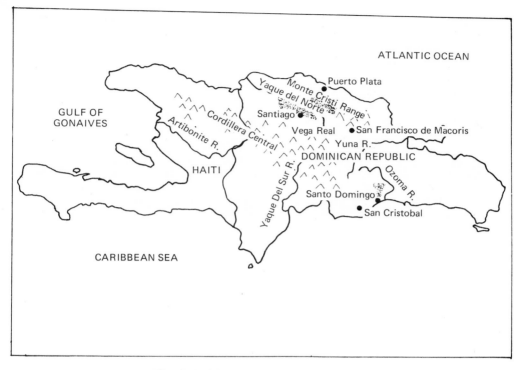

Fig. 7-1. Map of Dominican Republic.

Pedro Garcia in the Palo Alto de la Cumbre region, and the vertical outcrop here is approximately 300 meters in length. This mountainous location was opened for commercial mining by the government in 1949 under the direction of Dr. Pompilio Brouwer, director general de Minas y Petroleo, during the Trujillo adminstration, but large scale operations no longer exist.

These mountains consist mainly of Tertiary marine sandstones and shales, along with a variety of eroded conglomerates. Wherever carbonaceous material and lignite occur in the sandstones, amber is likely to be present.

In 1960, Sanderson and Farr [4] described a geological section of the mining area as follows:

. . . The uppermost layer [varies] in thickness up to 15 m. Below this layer in this section was a soft layer of clay shale varying from ½ to 2 m. in thickness, followed by a harder layer of silty shale 2 to 2.5 m. thick. Below the latter was a fourth layer of unknown thickness, of grey sandstone in which the amber occurred. Contacts between these units are gradational. The sandstone varies in color from light brown to dark grey, and it is a fine-grained, micaceous and carbonaceous and laminated gray wacke. The amber, which is confined to a thin bed, is removed by breaking chunks of sandstone, first removed with a pick, then by hand and with a heavy knife.*

* Reprinted with permission copyright 1960 by the American Association for the Advancement of science.

Fig. 7-2. Dominican miners examining lumps of amber in the dump in front of the mine opening in the Cibao region near Palo Alto. Soil is washed off the amber lumps before they are taken from the mining area *(Photo courtesy Giovanni Creations, San Juan, Puerto Rico.)*

Fig. 7-3. Raw amber lumps are transported to the workshops in the villages by pack mule. The amber is sold by weight. *(Photo courtesy Giovanni Creations, San Juan, Puerto Rico.)*

Fig. 7-4. Raw amber is sorted to select pieces with insect inclusions and clear, transparent pieces suitable for jewelry, and to discard unsuitable debris-filled pieces. *(Photo courtesy Giovanni Creations, San Juan, Puerto Rico.)*

Numerous small mines are scattered among the hills in the Northern Mountain range (Cordillera Septentrional) and are usually named after the small village located nearby; such as Palo Alto, where two qualities of amber are found, one appearing to be softer and perhaps younger than the other; Palo Quemado, where red amber is found associated with limonite (a reddish-orange iron oxide); La Toca, where a hard fine quality golden amber is found; La Valle, where the largest piece of amber ever found in the Dominican Republic was excavated; and Los Cacaos, where "blue" amber is mined from a strata composed of blue glauconite clay. Mines vary from shallow cauldron shaped diggings to narrow tunnels which follow a vein into the mountainside as far as 70 to 80 feet. Many mines in this region are almost inaccessable and can be reached only by hiking along steep mountain paths or by packburro.

Amber deposits running underground are mined using a pick to break loose the sandstone. Mining must be done without blasting or heavy pounding, as the impact would shatter amber nodules and cause fractures in larger lumps. The gray sandstone is brought to the surface in sacks by miners and spread on the ground in the sunlight for sorting. The "ore" is picked over and then flushed with water to remove waste sandstone and sediment. Next, the raw amber is transported by pack burro down the mountainside to workshops in the village where it is cut and polished.

The photographs (see Figs. 7-5 and 7-6) of the Giovanni Creations amber workshop in Santiago depict a typical Dominican factory where amber is graded according to size; then the large pieces cut to desired shapes using steel hand tools such as a hacksaw; and, finally, polished by using power-driven buffing wheels. The amber is found in pieces ranging in size from an

Fig. 7–5. Large lumps of amber are cut to size using steel tools. *(Photo courtesy Giovanni Creations, San Juan, Puerto Rico.)*

inch up to the size of a small grapefruit, and most pieces have a thick dull brown crust. Large pieces are rare and, of course, much higher priced. Porous lumps may soften after a few days of exposure to humid or moist atmosphere.

The largest natural lump of rough amber ever recovered in the Dominican Republic to date was from one of these small mines located in the mountainous Palo Alto region north of Santiago. The piece measures 18 inches across the largest portion and 7 inches thick and weighs a total of 18 pounds. Like most amber which is mined from sandstone locations, the rough surface is covered with a grayish crust. The gigantic amber specimen is now located in the Colón Gift Shop museum in Santo Domingo.

To the northeast of Santo Domingo in the Bayaguana basin a lowland region composed of clay and mudstone (lutite) also produces amber, but of a different quality than that from the Palo Alto mountainous region. The amber is very aromatic when burned and does not take as good a polish as the Palo Alto amber. Bayaguana amber tends to be soft and has a greasy luster. The first geological survey of this amber basin is currently in progress by Salvador Brouwer and will be presented at the 9th Annual Caribbean Geological Conference in August of 1980. The amber diggings are located along the streams of Sierra de Agua, where miners dig pits or wells about 15 ft. deep through sedimentary layers until reaching a horizontal black lignite layer which contains small amounts of amber. Under the dark layer is a

Fig. 7-6. Craftsmen in the Giovanni Creations Santiago workshop polish amber lumps using a cloth wheel on a motorized buffing machine. *Photo courtesy Giovanni Creations, San Juan, Puerto Rico.)*

grayish-white coarse grain sandstone and mudstone which also contains amber. One pit may produce from 4 to 6 pounds of amber ore and takes approximately two weeks to dig using hand tools such as picks and shovels.

It has been reported that recently amber was found by the Greater Antillies Ltd. while core drilling near Porta Plata on the northern coastal plain. The first layer of amber bearing earth was located at a depth of 40 to 43 ft. below the surface with a second layer at approximately 60 feet. The company spokesman dated the amber as coming mainly from a lower Miocene formation with the lower layer as upper Oligocene. The amber is associated with marine saltwater deposits and is dispersed in scattered pockets which are now mined using a simple open pit technique. When the amber lumps are first removed from the earth, they have a chalky white powdery outer crust which must be polished off by hand to display the transparent golden-brown amber.

The age of Dominican Republic amber is not definitely established. In 1959, Brouwer [5] tentatively assigned it to the Oligocene. However, recent studies by Dr. Francis Hueber, paleobotanist of the Smithsonian Institute, shows that the amber occurs in the Lower Miocene formations and is similar to that from Chiapas, Mexico. Specimens containing botanical inclusions suggest this amber is not an exudation from a coniferous tree, but from an ancient leguminous plant, possibly a species of *Hymenaea,* similar to the African variety which produces copal today.[6] Different qualities of amber occur in the Dominican Republic and Dr. Brouwer points out that these are in secondary deposits. Therefore, age of the amber can not be completely determined by the formation in which the amber occurs.

Dominican amber contains no succinic acid and therefore is classed as a retinite. The specific gravity is 1.048 to 1.08 and the hardness is 1.5 to 2—somewhat softer than Baltic amber. It polishes beautifully, but, because of its softness, care must be taken to prevent scratching or chipping. Dominican amber is generally less expensive than Baltic succinite.

Dominican amber ranges from transparent to opaque and its color varies from pale yellow to brown, with brownish yellow being the most common hue. Occasionally, deep reddish browns and fluorescent blues occur. Pieces of red and "blue" amber sell for 50 percent more than do the yellows and browns. Sinkankas[7] states that after a few hours of exposure to sun, red pieces may bleach to yellow or brown shades, although some pieces retain their original reddish hues. (See Color Plates 36 and 37 for various colors of Dominican amber specimens.)

An unusually fluorescent variety of Dominican amber is called "blue" amber, although in transmitted light, such pieces appear to be clear yellow to a yellow-brown color, and it is only in reflected light that a blue surface hue appears as if produced by an oily film. "Blue" amber is highly prized and commands a steep price on the market. When exposed to ultraviolet irradiation, all Dominican amber fluoresces intensely in blue and green shades. (See Color Plate No. 39 of fluorescent Dominican amber.)

Dominican amber contains abundant inclusions, from unidentifiable botanical debris to leaves and a wide variety of insects, but the insect inclusions are relatively scarce, it being estimated that only about one out of a hundred pieces contains an insect (see Figs. 7-7, 7-8 and 7-9).

Sanderson and Farr[8] identified several orders of insects, including Blattaria, Isoptera, Corrodentia, Heteroptera, Hymenoptera, Hemiptera, Coleoptera, Lepidoptera and Diptera. Also observed were spiders, fragments of wood, roots, flowers, leaves and air bubbles, with some of the bubbles of large size and some containing water which can be easily seen with a jeweler's loupe (10x).

The most recent study of fossils is currently in progress by Dr. Robert Woodruff, entomologist with the Department of Agriculture of Florida. Funded by a grant from the National Science Foundation, Woodruff is studying the Dominican amber fossils for evidence of the colonization of the islands through ancient ancestors of the modern species of insects. As a result of isolation, island insect species tend to evolve faster than continental that specific variety of fossil, the piece will receive a number, be identified and recorded, and then the specimen along with the information would be

Fig. 7-7. Dominican amber with a wasp inclusion (order *Hymenoptera,* family *Ichneumonoidea*).

species. To date Woodruff has found fossil representations of 19 of the 26 different orders of insects and is in the organizing stages of establishing a Central Clearing House for the international registration and identification agency for Dominican amber fossils. Tentative plans are that persons owning a piece of Dominican amber containing a fossil insect inclusion may send it to Woodruff for identification, he will direct it to the specialist in

Fig. 7-8. Large lump with flora and fauna inclusions and an unusual bubble filled with water. *(Courtesy Amber Forever, Florida.)*

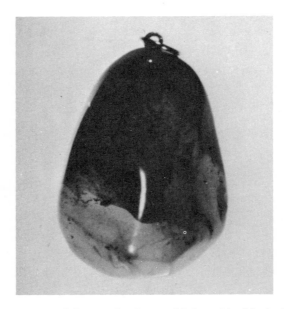

Fig. 7–9. Dominican amber lump with broad leaf inclusion.

that specific variety of fossil, the piece will receive a number, be identified and recorded, and then the specimen along with the information would be returned to the sender. When this agency is finalized, it will be of tremendous value to both fossil collectors by increasing the value of the specimen as well as benefiting the scientific studies in progress.

When amber mining was first begun by the Dominican government, much rough material was sold to German amber industries. Unconfirmed reports indicate that this material was used in producing ambroid or reconstructed amber items. Since Dominican amber reacts like Baltic amber to heat treatment, it is likely that similar pressed amber products could be produced.

It was soon evident more profits would be gained if local workers were trained in the art of polishing amber and finished jewelry were sold. Therefore, the Coindarte Foundation, or Cooperative of Industrial Artists, was organized in the early 1950's by Dr. Brouwer to provide a government supported trade school to teach amber working techniques. Emilio Perez, a local amber worker from Tamboril who was noted for the fine quality amber pipes and cigar holders which he carved, was the first instructor for the school in Santiago where local boys were given the opportunity to learn to polish amber. The government paid the original group of trainees 30 pesos per month while learning the trade. Soon special polishes for use on the delicate amber were imported from Germany and the cottage industry was begun. Currently there are many home artisans who have technical experience and are experts in this work.

Jewelry designed by the Dominican artisans tends to have a distinctive quality reflecting the Taino Indian culture of the past. In making jewelry, small nodules—called *marifinga* in the Dominican Republic—are sometimes tumbled just enough to remove the outer crust so that the amber is left in a semi-polished state. The nodules are drilled through and strung as bracelets and necklaces. Another common type of jewelry from the Dominican Republic uses gold-filled wire wrapped around each bead and requires no holes to be drilled. Transparent amber is generally used in articles constructed in this manner. The technique requires that amber pieces be shaped and polished on a felt wheel and grooves cut around the periphery of each bead to take the wire. The luminous quality of the Dominican amber is uniquely displayed in wire-wrapped jewelry employing beads of various shapes, including round, square, diamond and oval.

Small amber carvings are also constructed by local craftsmen and reflect native cultural influences. Pendants are typically carved to resemble a Haitian "tiki" god and for such purpose dark brown amber with much plant debris is generally used.

Another characteristic style of construction uses pieces of polished amber with small eye-pins inserted in each end, rather than drilled through as for beads. The loops on the eye-pins are then linked together to form a necklace.

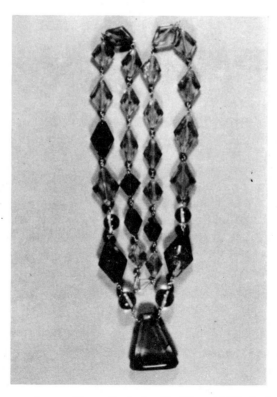

Fig. 7-10. Dominican amber necklace attached with 12k gold-filled wires. The center pendant contains two prehistoric insects. (Author's collection.)

Fig. 7–11. Dominican amber necklace attached with 12k gold-filled wires between each bead. The center pendant contains botanical inclusions. (P. Tice-Huff's collection.)

Today the Dominican Republic is the most plentiful source of amber outside of the Baltic area. Finished articles, fashioned locally by craftsmen in the growing cottage amber industry, as well as semi-finished lumps are currently imported into the United States. However, the local government has now restricted the exportation of completely unworked rough amber to encourage the development of an extensive cottage industry which uses one of the country's few natural resources. An unlimited quantity of amber articles can be brought into the United States duty free and semi-finished lumps with one side polished provide suitable and readily available materials for the lapidary desiring to cut and polish amber.

AMBER FROM BURMA: BURMITE

Burma amber, or burmite, has been known to Chinese craftsmen from as early as the Han dynasty (206 B.C. to 220 A.D.). Amber was brought from Burma through the Yunnan Province and it is possible that small amber deposits were found there, too. During a later period, Baltic amber was imported via the Mediterranean and India. Up until World War II, most burmite went to China.[9]

Amber was highly valued in China because it was thought to be a product of the pine, a tree held in particular reverence by the Chinese, who thought them to be symbolic of longevity. One of the favorite figures carved in amber by Chinese craftsmen was the fish, which also represented long life and good fortune. As early as the fifth century A.D., T'ao Hung-ching stated: ''There is an old saying that the resin of fir-trees sinks into the earth, and transforms itself [into amber] after a thousand years. When it is then burned, it still has the odor of fir-trees. There is also amber, in the midst of which there is a single bee, in shape and color like a living one. . . It may happen that bees are moistened by the fir-resin, and thus, as it falls down to the ground, are completely entrapped.''[10]   The Chinese also believed that after another thousand years, amber turned to jet.

Amber was known to the Chinese as ''hu-p'o,'' meaning ''soul of the tiger,'' as explained by Li Shih-chen:[11]

When a tiger dies, its soul [spirit] penetrates into the earth, and is a stone. This object resembles amber, and is therefore called hu-p'o [tiger's soul]. The ordinary character is combined with the radical yu [jewel], since it belongs to the class of jewels.

Being the soul of the tiger, amber was regarded by the Chinese as symbolic of courage and was supposed to possess the many strong qualities of the tiger.[12]

Burmese amber is highly prized for its deep colors and, in European and American markets, for its rarity. Usually, Burmese amber is a dark brown and largely somewhat turbid. Darker specimens verge on reddish shades, appearing deep red when intense light is transmitted through the mass. Peasants recognize 14 varieties of amber based on colors and shades. Cloudy varieties, typical of Baltic succinite, do not occur in Burmese amber, but an opaque variety, called ''root amber,'' is known. The various mixtures of brown shades characterizing ''root amber'' result from penetration of calcite into pores and openings in the amber, with cracks often filled with calcite in thin layers, the latter commonly found intersecting one another. This combination of amber and calcite gives burmite a curious mottled appearance.[13] In this connection, a variety of Baltic amber, also imported into China, is similar to Burmese ''root amber'' because of having become impregnated with salts from the soil and losing its color and transparency. When first found, it has a hard dark crust resembling an ordinary pebble, but upon polishing, it assumes a rich brown mottled appearance similar to marble.[14] Both kinds of ''root'' amber are used for carving ornaments with designs cleverly utilizing the natural swirls, color variations and actual flaws in the material to create remarkable effects. In China, this type of carving is called *ch'iao-tiao,* meaning ''clever or ingenious carving,'' and is, of course, also used in fashioning jade and hard stone ornaments.[15]

Burmese amber contains up to 2 percent succinic acid, much less than Baltic amber. Nevertheless, burmite is considered to be a succinite rather than a retinite. The chemical compositon of burmite is approximately as follows: carbon, 80.5 percent; hydrogen, 11.5 percent; oxygen, 8.43 percent; and sulfur, 0.02 percent. It is the hardest of all ambers (H = 2.5 to 3), and, though brittle, it is not difficult to work. If the piece does not contain too much calcite, it easily takes a beautiful polish. Its specific gravity of 1.034 to 1.095 and refractive index of 1.54 are similar to those of Baltic amber. Burmese amber is also highly fluorescent, with most varieties emitting a bluish or greenish color, while "root" amber pieces may fluoresce yellow-orange.[16]

Both botanical and insect inclusions are found in burmite. Recent botanical studies conducted by Jean Langenheim state that amber specimens from the Hukawng Valley in Burma proved to be from the Eocene and are related botanically to angiosperms rather than coniferous trees (family, *Burseraceae*).[17]

The rough is generally found as oval or elliptical lumps and occasionally in rounded pieces, but not in irregular or angular pieces. Large lumps were rarely obtained from shallow pits, as these tended to yield inferior amber. Large pieces were sometimes found in a finely-laminated blue sandstone or dark blue shale. The largest lump ever found was the size of a man's head, but most pieces were small, flat and smooth.[18] The amber occurs in shales and sandstones which are alternately interbedded with layers of limestone and conglomerate. Carbonaceous impressions and, infrequently, very thin coal seams containing concretions of amber, are embedded in the sandstones and shales. Since both are soft, surface exposures are rare. In sections along streams, where erosion has taken place, pits and cuttings expose the earth's structure and reveal the true nature of its geology.[19]

The first geologist to visit Burma amber mines was Noetling, in 1892. He concluded that the deposits occurred in clay and wrongly ascribed a Miocene age to them. Based on this report, Bauer's *Precious Stones* described amber lumps as "distributed sporadically throughout a bluish-grey clay belonging to the lower Miocene division of the Tertiary formation."[20] However, in 1922, Murray Stuart visited the mines and found occurrences of *Nummalites biarritzensis*, an index fossil for the Eocene epoch, among shales embedding amber. An Eocene age was also assigned in the work of F.A. Bather and T. Cockerel, based on studies of insect inclusions. In 1934, H. L. Chhibber, geologist and author of *Mineral Resources of Burma*, visited the amber mines and conducted an extensive survey which confirmed the Burmese amber deposits belonged to the Eocene.[21] (See the map of Burma in Fig. 7-12.)

Chhibber provides a complete description of primitive amber mine workings in the Nangtoimow Hills in northern Burma, the most famous mines being near the jadeite mining area. Approximately 200 pits at that time were located about 3 miles south of the village of Shingban near a small stream between Shingban and Noije Bam. Other open pits were operating nearby in Pangmamaw at the time of Chhibber's visit.[22]

The largest amber mining center was located at Khanjamaw, about 3 miles

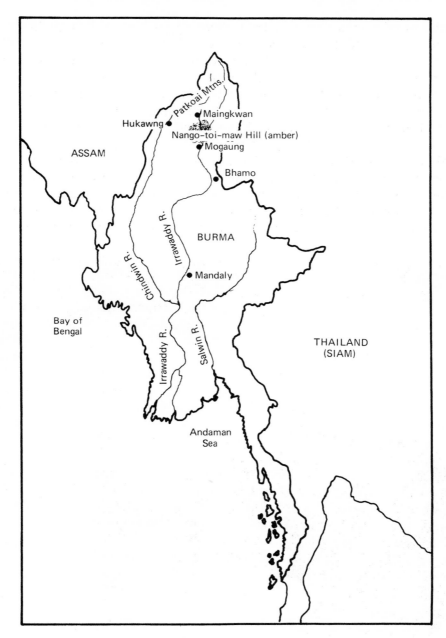

Fig. 7-12. Map of Burma.

southwest of Shingban and supporting about 150 miners. Workings consisted of pits lined with thin bamboos supported by heavy wooden posts. The deepest pit had been dug about 45 feet deep, to a point just above the water table level, which appeared at a depth of 49 Feet. At a depth of about 40 feet below the surface, coal seams containing good amber were encountered.

Another small working, called Ningkundup, was located about 200 yards northwest of Khanjamaw. In 1934, there were 6 pits worked by 15 Shan-Chinese. The strata here consisted of blue sandstones and dark blue shales. At Wayutmaw, 20 miners, working in small teams with primitive tools and bamboo hoists to raise baskets filled with soil, were engaged in redigging old pits. This mine was next in importance to Khanjamaw. A little northwest of Wayutmaw, where amber-bearing strata consisted of bluish sandstones and finely bedded shales covered with a reddish overburden, the amber was washed out by sluicing during the rainy season.

About 700 meters southwest of Khanjamaw, the Chetauk mines consisted of seven pits. However, about 500 to 600 pits, scattered in the neighborhood, marked places where mining was abandoned when no more amber could be found. Other, less important, mines were located in Ladumnaw and Lajamaw, where amber occurred at a depth of approximately 24 to 27

Fig. 7–13. Shan-Chinese working an amber pit using levers for raising debris from the mine (Hukawng Valley). (From Chhibber, *Mining in Burma,* 1930.)

feet. The workings were near small streams, and pits were often filled with water. Mining had to be carried out during the dry season, which lasted about five months. It is obvious that the same amber-bearing strata were worked in these groups of scattered, small pits.

According to Chhibber:[23]

The amber mines are really shallow wells about three feet six inches square with a maximum depth of about 45 feet. The number of coolies working a pit varies from two to four, but generally three men work together. One of them digs underground with a hoe (wooden crowbar tipped with iron), while the second one hands up the debris in small bamboo baskets, either by means of a Madras lever or by a long bamboo with a bent hook at one end. The soil is then looked over for amber nodules. The third miner dumps the soil. If four of them work together, then two of them work underground. The deep pits, especially when they are largely in the soft shales and close to one another, are lined with thin bamboo barricades. In deep pits water appears at a depth of about 40 feet, and is bailed out by hand methods; as water level is, of course, higher in the low ground. Mining is continued until a layer of sand is reached.

In 1904, not more than 300 pounds (136,08 kilograms) of amber per year was produced, as the mine was operated by peasants with very primitive

Fig. 7-14. Damp amber mining pits were lined with bamboo to support the insides of the shaft. The man is sitting in a primitive bamboo pump arrangement which helped to remove water from the pit (Nangtoimow Hills). (From Chhibber, *Mining in Burma,* 1930.)

Table 7-1. Output of amber 1926 to 1930.[24]

Year	Quantity (lb)	(kg)
1926	3950	1791.32
1927	7050	3197.8
1928	2940	1333.55
1929	1958.5	888.35
1930	207.3	94.03

tools. Little improvement in mining methods had taken place by the time of the geological survey conducted by Chhibber in the 1930's. Production reached its peak in 1927 and declined each year thereafter (see Table 7-1).

Some amber was polished in the mining area using simple cutting tools such as a saw, a *Kachin dah* (a large knife), a file and sandpaper. First the crude amber was cut into small pieces with a saw made by fixing a piece of kerosene tin into a bamboo handle and cutting teeth in the blade with a *dah*. Using this same knife, rough pieces were cut into shape, and shaping was completed with small flat files. Each piece was then sanded by rubbing with

Fig. 7-15. Luminous transparent yellow amber from Thailand (Siam). Beads are carved in Chinese baroque style. (Author's collection.)

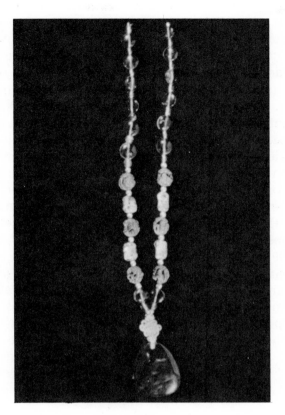

Fig. 7-16. Tabular-shaped Burmese amber beads and pendant combined with carved peach pits and root amber beads. (Author's collection.)

course "glass-paper." The final polish was obtained by using a tree leaf with a prickly-backed surface.[25]

The major portion of production was sent to Mandalay and Mongaung. Again, very simple tools were used to fashion long round-beaded necklaces for Mandarins. Amber was also cut into beads for necklaces, earrings, bars for brooches, buttons, cufflinks and other trinkets. However, so many other articles were made that such adornments represented only a small fraction of the amount of amber used in the Kachin Hills, Hukawng, Naga and Chin Hills. In Maingkwan, large amounts of amber were used for making of *nadaungs,* or Chinese style earrings. Important centers for amber trade were the Maingkwan, Mogaung, Mandalay and the Naga country of Assam.

The symbolic value placed on amber led to much amber being imported from Baltic sources, as well as from other Asian countries where amber is produced. A particularly luminous, clear, pale yellow variety was imported from Siam (Thailand) and is remarkable for its light sherry or golden color, similar to that of Baltic amber, and its usual uniformity in color. This amber was called *Ching Peh,* meaning golden amber, and was often worked into round or baroque shapes.

In the traditional necklaces of the Mandarins, all beads were required to be uniform in size, to be round and to number exactly 108 beads because each bead had its own religious significance. Another shape of amber bead from China is the flat, tabular type. (See Fig. 7–16)

(Color Plate 40 shows transparent brown amber beads with a darker brown turbid shading. The striped appearance in this variety of amber led to its being called *hu peh,* or tiger amber. Also included in this necklace are four beads of the variety called "root" amber. In Color Plate 41, a typical round beaded amber necklace is shown, including jade ornaments. This style was worn for special occasions, such as an audience with a high offical.)

Two other varieties of amber received special names from the Chinese. A translucent yellow amber was called *mi la,* or honey amber, whereas a darker yellow, opaque variety was called *Chio-Naio,* or bird's brain.[26] Another amber article highly prized by collectors today is the antique snuff bottle. Some of these are exquisitely carved, but others are plain, allowing the beauty of amber to speak for itself.

Because of the high value placed on amber, it was often counterfeited in China. As early as the sixth century A.D., T'ao Hungching (452 A.D. to 536 A.D.)[27] warned that "only that kind which, when rubbed with the palm of the hand, and thus made warm, attracts mustard-seed, is genuine." The electrostatic property of amber was often used to distinguish it from imitations, but in about 1910, bakelite plastic imitations reproduced even this quality, making identification more difficult. Recently, large carvings made

Fig. 7–17. Close-up of "hu peh", or tiger amber, to show the striped appearance of the amber.

Fig. 7-18. Pressed amber was used to form blocks large enough for carving; however, today, plastic materials are being used to produce these large red art objects.

from a transparent red synthetic resin are appearing on the market and are claimed to be "reconstructed Burmese amber," but obviously they are much too large to be true amber. However, to produce blocks large enough to be carved, pressed amber has been used in China for many years. During the manufacturing process, amber chips are mixed with linseed oil and a red coloring added. Nevertheless, the majority of such large carved red art objects are constructed of a synthetic resin similar to bakelite plastic. The only way to be sure any piece of this type is genuine is to test it as explained in the chapter on testing and imitations.

Very little Burmese amber reaches markets outside China today. For the collector looking for Burmese amber at the present time, the best place to search is in antique shops or from importers dealing in Oriental antique ornaments. In 1968, Sinkankas reported in *Van Nostrand's Standard Catalogue of Gems*[28] that "Burmese amber is rarely available today."

RUMANITE, AMBER OF ROMANIA

Amber from Romania was named rumanite in 1891 by Otto Helm.[29] It reached its height in popularity during the early 1900's, but is very rarely seen today. Besides being used for amber jewelry, rumanite was sent to Vienna to be manufactured into pipe stems and other smoker's supplies.

Rumanite was prized for its rich colors, which range from brownish yellow to deep brown, with deep colors predominating. Colors such as dark garnet, smoky gray and greenish to almost black are not unknown. It may have a mottled appearance resulting from veins and cracks which reflect a metal-like sparkle.[30]

Rumanite does not include the opaque, bone-like varieties so commonly found among Baltic ambers. More often it is transparent, with an opaque crust that is darker in color than the interior of the mass. A variety of rumanite referred to as "black amber" is actually a deep ruby red, blue or brown when held to a light source allowing light to be transmitted through the resin. The term "black amber" is misleading, as there is no truly black amber and none was used by the jewelry industry. Stantienite is a black fossil resin, but it has not been used commercially. However, another substance often called "black amber," and sometimes confused with amber, is in reality jet. Jet is a variety of lignite coal, a fossil wood similar to cannel coal.

The deep colors of rumanite are thought to be influenced by sulfur deposits which occur in the same places where amber is found. The gases absorbed in the amber also tend to discolor it. In addition, inclusions of hydrocarbons (such as coal) or minerals (such as pyrite) may occur, and each contributes to the darkening. It is thought that the presence of such inclusions also contributes to the opalescence of the surface. Pieces of brownish red amber occasionally reflect a blue-green fluorescence, similar to that of the "blue" amber from the Dominican Republic. Rumanite is highly fluorescent under ultraviolet light.

A significant characteristic of rumanite is its sulfur content, which amounts to as much as 1.15 percent. When burned, the choking fumes of hydrogen sulfide are readily detected. Rumanite fuses at 300 to 310 degrees centigrade without swelling or altering shape. The melting point is 330 degrees centigrade as compared to the 287 degrees centigrade for Baltic amber.[31]

Rumanite is similar to Baltic succinite in structure and breaks with a conchoidal fracture. Compared to the density of Baltic amber at 1.05, rumanite is 1.048. The refractive index of rumanite is 1.438, compared to that of Baltic amber at 1.54. Rumanite is also slightly harder than Baltic succinite.

A chemical analysis of rumanite indicates that it contains 1 to 5 percent succinic acid, less than the amount found in Baltic amber. It is classed as a succinite, and offers more resistance to solvents than does Baltic succinite.[32] (See Table 7–2 for solubility.)

Rumanite occurs in a Miocene formation, and thus differs from Baltic amber, which is found mainly in Upper Eocene/Lower Oligocene deposits. Hydrocarbon studies, using infrared spectrophotometry and X-ray diffraction patterns of crystalline constituents, indicate it may be of a leguminous rather than coniferous origin.[33]

Rumanite was known and used as a gem by the Romans, who established

Table 7–2. Rumanite solubility.*

Solvent	Percent of Solubility
Alcohol	6
Ether	16
Chloroform	10
Benzene	14

*These figures vary with different specimens. Percents from Williamson.

colonies in the country and gave it its name. The Bužau area ruins yielding Roman artifacts also included rumanite objects, proving that the amber was known at that time. Furthermore, Roman trade routes—which include Klaipeda (Memel), Lithuania, in the center of the Baltic amber area—pass through Bužau, one of the localities where rumanite is found.[34]

The first written work mentioning the amber of Romania was published in 1822 by Zuicker, the German geologist who first investigated the area's geological history and studied the resin from the banks of the river Bužau. It was he who placed the age of rumanite in the Miocene epoch while studying the strata of sandstone near the river where amber was found. The Bužau deposits, the main source in the early 1900's were located along the railroad lines between Bucharest and Braila.[35,36] (See Fig. 7–19.)

Rumanite is mined not only from sandstone near the Bužau River, but is also found in the Carpathian Mountain region. Recent Soviet reports on prospects of amber in USSR state that the Carpathian area where rumanite and other fossil resins are found covers an area of Ciscarpathia in the Soviet Carpathians near the upper basins of the Prut and Dniester (Dnestr) rivers as far as Khatin (Khotin). The amber is said to be suitable for use in jewelry.[37]

Other deposits are located near the Danube, in tar deposits of Putna, and along the banks of brooks near Valeng de Munty (Valeni de Munte). Rumanite is reported elsewhere in Romania at Ramnicul-Serat and Prahova and along the shores of the Black Sea in the province of Dobrogea, also near

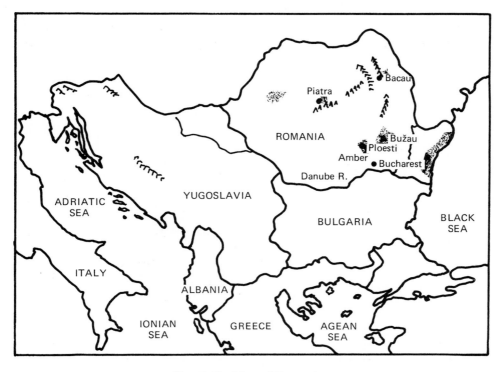

Fig. 7–19. Map of Romania.

Valzea in the province of Oltenia, and at Bacau, Meamt, and Sucava in Moldavia. A yellow amber is found near Sibiu and Alba in Transylvania.

None of this amber is currently reaching the United States. During the late nineteenth and early twentieth centuries, the only commercial rumanite came from the Bužau area. Although amber from the Carpathian mountains is still of interest to the jewelry industry, it is not produced on a commercial scale. The most likely source of rumanite jewelry today is from old collections, since rumanite is by far the least plentiful type of amber. In this connection, the only rumanite this writer was able to locate is preserved in the Field Museum of Natural History, Chicago, Illinois, and in the Cranbrook Science and Mineral Museum, Bloomfield Hills, Michigan. Several old faceted beaded necklaces, circa 1920 to 1930, sold by antique dealers as genuine rumanite, proved to be bakelite.

SIMETITE, AMBER OF SICILY

One of the most beautiful and highly prized varieties of amber occurs in Sicily and is named simetite after the Simeto River on the eastern coast of Sicily near Catania. The mouth of the river was the major source but some was also found near brooks and streams in clay deposits in the central part of the island. The entire area around Etna and Syracuse is volcanic, and amber found within this area possesses a special opalescence and luminescence. It has been thought by some that the volcanic activity is the cause of this unusually beautiful coloring.

Simetite amber is rarely yellow, but more often a transparent dark red, blue or green. Red specimens are highly prized and were regarded as the

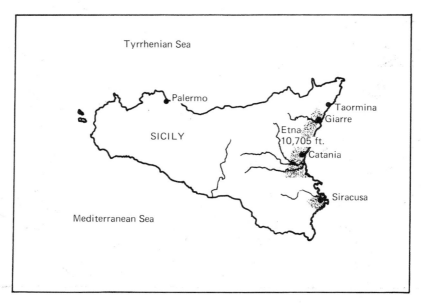

Fig. 7-20. Map of Sicily.

most desirable variety in the late 1800's. Expounding on the radiant beauty of the "Ambra de Sicilia," in 1897, Arnold Buffum wrote:

. . . the gems in her necklace flashed in the sunlight, showing color shades ranging from faint blue to deepest azure, from pale rose to intense, pigeon blood, ruby red. The varied and lustrous hues, here blended in lavish beauty, drew from me involuntary expressions of admiration.[38]

Buffum emphasized that he could not believe the hues were natural, but when assured they were "pristine" hues, he exclaimed, "they are hues, then of the primeval world, the imprisoned color shades of an earlier and more exuberant clime. . ."[39]

THE

TEARS OF THE HELIADES

OR

AMBER AS A GEM

BY

W. ARNOLD BUFFUM

WITH ILLUSTRATIONS

LONDON

SAMPSON LOW, MARSTON & COMPANY, LIMITED

St. Dunstan's House

FETTER LANE, FLEET STREET, E.C.

1896

Fig. 7-21. Title page of Buffum, 1896

By the turn of the century, amber in such beautiful colors was scarce in Sicily. The widespread, isolated, nest-like surface deposits were already greatly diminished. Buffum cautioned: "it would not be surprising if these deposits were to cease and the amber of Sicily to disappear."[40]

The refractive index and specific gravity of simetite are similar to those of Baltic succinite. Its hardness is 2.5 on Mohs' scale, and it has the same conchoidal fracture and electrical quality as Baltic succinite.[41] Simetite contains no succinic acid and is classified as a retinite. When burned, the fumes are not irritating; however, along with the resinous odor, a sulfurous odor is emitted, resulting from the relatively high sulfur content (which varies between 0.67 and 2.46 percent). Red specimens contain smaller amounts of sulfur, with the deeper brown shades containing the larger amounts. Simetite is highly fluorescent under an ultraviolet light, as well as in reflected daylight.

Modern botanical investigations indicate that the simetite resin source-tree is related to the *Burseraceae protium,* an angiosperm, rather than a conifer.[42]

Simetite generally occurs in Tertiary lignite, dated to the Middle Miocene. A black resin is sometimes associated with simetite, occasionally forming a coating over simetite nodules, but this is believed to be of different origin and has no commercial use.[43]

Disagreements have developed over how long the existence of Sicilian amber has been known. Buffum presented the argument that Phoenicians were aware of it. Not only were they great traders, according to Buffum, but they also knew that to sell their wares for the highest price they needed to guard their secret sources of supply. One report of a Phoenician shipmaster

Fig. 7-22. Specimen of Sicilian amber.

tells of his running his ship aground—wrecking not only his, but the Roman ship following him—and thus preventing the Romans from learning the secret of his route. For this deed of valor and shrewdness, he was rewarded by his state. Buffum believed that the Phoenicians traded with the inhabitants of "Trinacria," the Phoenician name for Sicily, perhaps receiving amber, but that they took drastic steps to keep their trade secrets, telling tales of incredible dangers and monsters in that district.

Other authorities doubt that trade in simetite existed in this early period, since amber found in tombs excavated along Phoenician trade routes generally proved to be succinite from the Baltic. Such tomb amber appeared red when found, but upon cutting, it displayed a yellow inner core typical of Baltic amber.

Ornaments or jewelry made of simetite are rare, and most examples can be found only in museum collections. Local craftsmen of Taormina occasionally offer articles carved of Sicilian amber, and antique collectors are constantly alert for red varieties when viewing old estate jewelry collections, since this amber was more prevalent in the past. Unfortunately, red Sicilian amber was often expertly imitated, especially in the synthetic bakelite plastic, which provides a beautiful imitation with many properties similar to those of amber, including some of its electrical powers. Bakelite is often passed off as "cherry red amber." However, since bakelite is much heavier, more uniform in texture, clearer and harder than amber, it is not difficult to identify. For these reasons, red "antique amber" should be tested for refractive index and specific gravity.

AMBER OF MEXICO

Mexican amber is found near San Cristobal las Casas in Chiapas, in the form of lumps large enough to be polished. Although the Chiapas amber has only recently been publicized, it was well known to the area's ancient inhabitants. It is said that Montezuma used a long amber ladle, carved and polished by the skilled jewelers of Xochimilco, to stir his cocoa drinks.[44]

Chiapas amber occurs in alluvium in the Simojovel district, where heavily forested mountains overlook the Huitapan River. After heavy rains, amber is washed out from the sides of gorges and carried downstream, where it is collected by the natives.

Fran Blom, archaeologist and anthropologist, spent several years studying the Simojovel region where amber is obtained and also found amber near Yajalon, at a location called Hool Babuchil, or "Head of Amber," a deposit exposed by a steep fault. Amber is found along the surface in horizontal strata, as well as at the foot of the cliff, where it has weathered from soft marine limestone often containing shell fossils. This deposit was studied in 1954 by two scientists from the University of California, J. Wyatt Durham, paleontologist, and Dr. Paul Hurd, entomologist, who collected many specimens, later taken to Berkeley for further study.[45]

Chiapas amber is similar to Baltic in refractive index and specific gravity, but is not as hard, and occurs in various shades of red, yellow and almost

Fig. 7–23. Broad leaf inclusion in amber from Chiapas, Mexico. (Autnor's collection.)

black. The Natural History Museum in Nuevo Chapultepec Park in Mexico City contains an exhibit of five pieces of Chiapas amber from Fran Blom, all being reddish brown and ranging from the size of a walnut to that of a small fist.

Chiapas amber was found to be a fossil resin formed during the Miocene and/or Oligocene epochs of the Tertiary. The source of this resin is thought to be a leguminous tree and it is classed as a retinite rather than a succinite.

Locally, natives collect amber for wear as amulets, or make small trinkets, carvings and pendants to sell to tourists. The amber is cut and shaped with saws and steel files and polished by rubbing with a mixture of water and ashes on a soft piece of balsa-like wood.[46] However, many imitations are made from a recent gum, which is bright yellow in color. When held in the hand, such imitations become sticky to the touch. Being recent in origin, their surfaces do not keep a polish but become crazed and lose luster. When burned, such gums will drip, whereas amber burns and emits a fragrant resinous odor along with a dark smoke.

Baja California has also been the source of a fossil resin which has received much study by scientists. Insect inclusions were found in the amber from this area. One piece included a whitish bee-like insect and a grub.

CANADIAN AMBER: CHEMAWINITE OR CEDARITE

Canadian amber has been known for over a century, but little has been collected or used in ornament. The best known and most widely studied source is Cedar Lake, Manitoba. In 1889, local Indians brought specimens to W. C.

King, the officer in charge of the Chemawin Trading Post. J. R. Tyrrell, of the Geological Survey of Canada, also obtained and studied the amber, while surveying the northern Manitoba area. This beautiful amber or fossil resin is named Chemawinite after the Indian name for Hudson Bay Company post, which is near the deposits.[47] Though Canadian amber is most abundant at Cedar Lake, where it is found in alluvial deposits, it also occurs elsewhere in low grade coal and lignite beds and in carbonaceous sediments.

Chemawinite is found as small nodules of brownish color in sands and gravels along the southwest edge of the Cedar Lake and is considered to be Cretaceous in age, 60 to 130 million years old. It contains no succinic acid and therefore is classed as a retinite. The source trees are not known, but Dr. Jean Langenheim relates the resin to that from the Araucariaceae.[48]

In 1891, O. J. Klotz, of the Geological Survey of Canada, attempted to find other deposits in the Cedar Lake region, but results were disappointing.[49] Only traces of amber were found at several places along the Saskatchewan River flowing into Cedar Lake, and none of the deposits were rich enough to support a commercial mining venture. In 1933, however, the Craig Amber Mine at Cedar Lake supplied a small quantity of gem material, but in all, only about one ton of amber was produced from this area between 1895 and 1937. In still another attempt to exploit the deposits, the firm of Native Minerals Limited attempted to obtain amber from the region for commercial purposes in the mid-1950's. During the winter, amber-bearing sand was scraped up from the shore and shipped to Winnipeg for sorting. This venture also proved to be unprofitable because the amber was found mainly in the form of very small granules. Some of these specimens were donated to the Canadian National Collection of Insects.

Chemawinite is important scientifically for its inclusions of well-preserved insects, spiders and mites, among the best to be found anywhere. It also contains pollen grains, spores and fragments of plants from the Upper Cretaceous period, some 70,000,000 years ago, as well as many unusual insects, which, though predominant during this epoch, are now extinct.

In 1937, R. M. Carpenter,[50,51] a paleontologist of Harvard University, organized an expedition to Cedar Lake to collect Chemawinite inclusion specimens. Working for three months, he obtained 400 pounds (181.4 kilograms) of amber. Later attempts to collect specimens for scientific study yielded smaller and smaller quantities. In 1950 and in 1963, for example, W. J. Brown and R. D. Bird, collecting for the Canadian National Collection of Insects, found only 2.5 pounds (62 grams) of amber nodules, ranging from the size of a pea to that of a robin's egg. The amber was found along the sandy shores of the Cedar Lake mouth of the Saskatchewan River, mingled with shells, coal fragments and organic debris. In 1969, McAlpine and Martin reported the site no longer accessible because a dam at the foot of Cedar Lake at Grand Rapids had raised the water level to cover the amber-collecting areas.[52]

During their studies of inclusions in Chemawinite, McAlpine and Martin examined 470 pieces with identifiable inclusions of fauna. About 300 different species were represented, but only 30 of these were completely

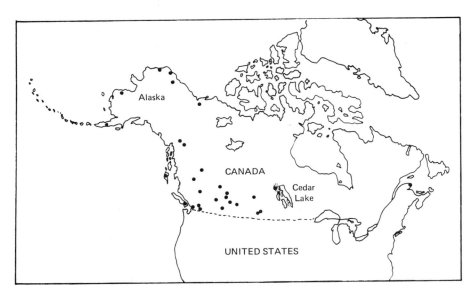

Fig. 7-24. Map of Canadian amber locations.

described at that time. Interestingly, the specimens studied did not readily fit into present orders, as is the case with those species found in Baltic amber, which can be assigned to modern genera or related categories.

Too little is known about Cretaceous amber insects to draw conclusions about their paleoecology and environment, but Canadian amber provides certain useful clues. First, no species—and very few genera—occurring in Cretaceous amber occur either in Baltic amber or any other place in the world today, which suggests that these evolutionary lines suffered extinction either in late Cretaceous or early Tertiary times. Second, some species found in Canadian amber, namely aphids, may be directly associated with coniferous trees. Since 40 percent of the inclusions in Canadian amber are Diptera, a rich, moist environment is indicated. The arthropod fauna suggests climatic and ecological conditions similar to present day Florida. The study by McAlpine and Martin revealed that the species in Canadian amber are more closely allied to species now living in Australia, New Zealand and South Africa than to any others.[53]

Other than the Cedar Lake occurrence, there are thirty-two places in Canada where amber has been reported, none of which are significant producers. These locations are shown on the map in Fig. 7-24.

AMBER AND FOSSIL RESINS IN THE UNITED STATES

Amber and fossil resins are found in small quantities in coal and lignite beds in various places along the Eastern Sea Board and Gulf Coastal Plain of the

United States, as well as in California and Alaska. Most of these deposits are small and cannot produce amber on a commercial scale. Some locations received considerable study by geologists and botanists and are mainly of scientific interest.

Arkansas

Amber in the Gulf Coastal Plain was first reported in 1858 by D. D. Owen in the lignites of Poinsett County, Arkansas. In 1860, Owen reported small pieces of amber-like resin found in lignite near Camden, Ouachita County. In 1959, Sinkankas described amber pieces up to 3 inches by 3 inches by 1 inch in Hot Springs County near Gifford,[54] but the deposits were poor and amber is no longer collected in the area. In 1971, amber was found near Malvern, Hot Springs County, Arkansas. R. Mapes and G. Mapes[55] recorded the collection of 300 grams of amber containing several insect inclusions, while others found more than 8 kilograms of amber in the lignitic clay of Eocene strata. More than 900 insects, arachnids and plant inclusions were isolated and studied. The amber pieces were generally small (3 to 8 centimeters) and ranged in color from clear, pale yellow to dark brown-orange. Numerous impurities, marked by dark opaque bands within translucent areas, are present, and only a very small portion is of gem quality. A representative Arkansas amber collection is preserved in the Museum of Comparative Zoology, Harvard University.[56]

Fig. 7-25. Amber nodules collected from lignite clay near Melvern in Hot Springs County, Arkansas. Some specimens clearly contain bark inclusions. (courtesy M. Berry)

Texas

Amber has been found in Cretaceous coals of Maverick and Brewster counties. Some pieces were translucent and of a quality suitable for lapidary purposes, but they were much too small. A poor quality brown amber has been found in Tertiary deposits along the Gulf Coastal Plain, but it is not gem material.[57]

Tennessee

Small pellets of Cretaceous amber occur in beds of lignite north of Newman Cemetery in the northwest section of Hardin County near Coffee Bluff.[58] The first known insect discovered in North American amber was found in 1917 in a lump of amber from this locality and was identified as a caddis fly.[59]

New Mexico

Fossil resins occur in lens-like sheets filling fractures in several coal deposits, but the material is small in size and none is of gem quality. However, jet is found and is polished by the local Indians and hobbyists.[60]

North Carolina

Small pellets of Cretaceous amber occur in lignite beds along a cut-off of the Neuse River near Goldsboro in Wayne County. The amber is generally a clear yellow color and is transparent, but only a very small portion is suitable for lapidary work. Local hobbyists collect buckets of the soil along the river, then wash it to remove the heavier materials and separate out the small amber pellets. Because only small quantities are found, the amber is not used commercially.

In other areas along the eastern part of the state, especially in Pitt County, amber is found in marl beds.

Maryland

Gem hunter's literature frequently mentions amber in Anne Arundel County, Maryland, where it was found over a century ago in the form of small pellets dispersed in sands of the Magothy Formation of Upper Cretaceous age. The locality, North Ferry Point, is no longer productive.

Massachusetts

Before 1883, a large 12-ounce (340.2-gram) specimen of fossil resin was found on Nantucket Island in a Tertiary greensand and marl formation. Smaller samples from the area were described as containing ants and a fly.[61] Today, small nodules of yellow amber may occasionally be picked up on beaches on the western point of Martha's Vineyard, but most are too small to be gem quality.

New Jersey

The largest quantity of amber from any area in the United States was found in the late 1800's in Monmouth County in Cretaceous formations. Subjected to a variety of tests, it was found to differ from Baltic amber, having a higher specific gravity and a tendency when heated to liquify rather than to soften and fuse. Because most of the material was found in marl (fertilizer) pits that no longer are worked, very little is now found.[62]

Kansas

A dark amber occurs in lignite beds along the Smoky Hill River in Elsworth County, but the beds are no longer accessible because of being covered by waters impounded by the Kanapolis Dam.

New York

Amber was found many years ago near Kreischerville on Staten Island, New York City, but according to the Buffalo Society of Natural Sciences, amber is no longer found anywhere in New York State.

California

Brown and honey-yellow ambers, along with lignitic wood, occur in clay shales in Eocene rocks in California. Ambers found in Simi Valley of Ventura County contain succinic acid, and Langenheim,[63] while studying botanical affinities of this resin to recent trees, related the resin to that produced by the *Pinaceae* (Pinus). California fossil resin appears to be slightly older than Baltic amber.

Other areas and exact descriptions of locations for sources of amber in the United States are reported in *Gemstones of North America*[64] and *Gem Hunter's Guide*.[65] Most of these areas produce merely small flecks of amber in coal or lignite formations and the amber is not of value as a gem. If a specimen is desired, these books may help in locating places where collectors can search. (See Fig. 7–26 for amber locations in the United States.)

AMBER IN OTHER AREAS

Japan

In about 1958, geological engineers searching for coal uncovered important amber deposits in Fushan from which, for a time, approximately four tons per day were mined. The material was used in Japan for making lacquer and none was exported.

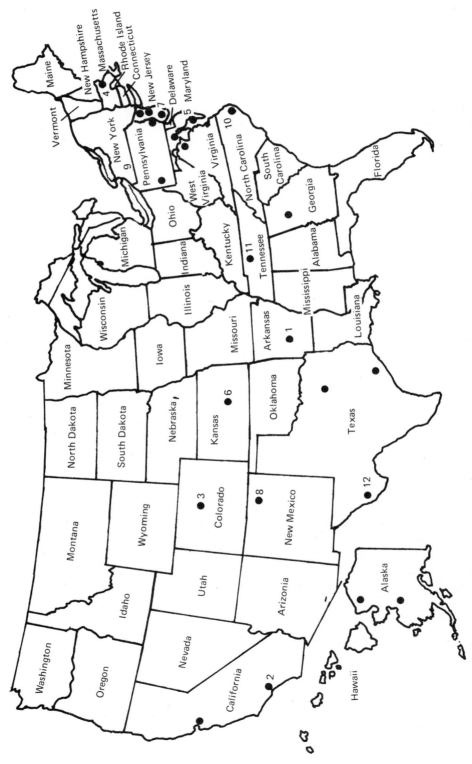

Fig. 7-26. Amber locations in the U.S.A. (From *Gem Hunter's Guide and Gemstones of North America*).

Amber Sources in the United States

From Gem Hunter's Guide and
      Gemstones of North America

1. Arkansas—Found in the South central area of the State. Gifford, Hot Springs County, in lignite clay near Missouri Pacific Railroad.
2. California—Simi Valley, Ventura County, found with lignitic wood in Eocene rocks.
3. Colorado—Boulder County and adjacent counties, in the coal of the Laramie formation.
4. Massachusetts—Gay Head, Dukes County, on the western tip of Martha's Vineyard. Found in Tertiary greensand and marl formations. In 1883 an unusual specimen weighing 12 ounces was found on Nantucket Island.
5. Maryland—North Ferry Point, Anne Arundel County on the Magothy River, Amber is gone, however. Currently Sullivan Cove on the Severn River on the north shore is a better location.
6. Kanasas—Ellsworth County, but the Kanopolis Dam has covered the area with water.
7. New Jersey—Neptune City, Monmouth County, in the marl along the Shark River. In 1886 this area produced the largest quantity of amber in the United States,. Today very little is found.
8. New Mexico—Raton, Colfax County found in the Sugarite Mine and in the Yankee Coal Bed, in lumps and streaks.
9. New York—Staten Island Corough of Richmond, New York City. Specimens found many years ago, and no longer available.
10. North Carolina—Pitt County and in the eastern part of the state, found in marl beds.
11. Tennessee—Tennessee River, Northern Hardin County, Cretaceous age amber found in the sands of Coffee Bluff.
12. Texas—Terlingua, Brewster County, brown and yellow amber found in Cretaceous coal at Eagle Pass and along the Terlingua Creek. Plentiful but very small nodules.

Sinkankas, Gemstones of North America p. 596–603
      Gem Hunter's Guide, p. 110, 177, 175, 204, 210.

Tanzania

About 1976, a find of so-called "young amber" was made somewhere in this country and became available on the United States market. It is reported by the Geological Enterprises Company to be from a Pleistocene formation, which would make it older than copal resins, but much younger than Baltic amber. It is transparent, pale champagne color, and softer and more brittle than Baltic amber, but with similar refractive index and specific gravity. Small insect inclusions are present, and the crust often exhibits impressions of leaf veins and cells. The material polishes well, but because of its softness, it will not take much hard wear. (See Color Plate 42.)

New Zealand

New Zealand is the source of another true fossil resin called *ambrite,* which is transparent and yellow in color. It has been known for many years but does not appear to have been used to any great extent.

Fig. 7-27. Polished pendent of Tanzanian "young amber" (Pleistocene).

OTHER FOSSIL RESINS

Many fossil resins are popularly, if misleadingly, called "amber," but in the past, mineralogists and gemologists referred to Baltic amber or succinite as "true amber," since it was most abundant and more often used as a gemstone. More recent studies by botanists indicate that no one fossil resin is a "true" amber. More realistically, fossilized resins may be termed either retinite or succinite, and deserve the name amber. Several other fossil resins have been known since mining of Baltic amber began, and some commonly

Fig. 7-28. Raw piece of Tanzanian amber with impression of leaf veins and cells on the surface.

occur along with Baltic succinite. Traditionally, these fossilized resins were compared to Baltic succinite when considering them from the gemological point of view. However, most of the following fossil resins are not fully suitable for use in the jewelry industry and in the past have been called "pseudo-ambers."

Gedanite, a fossil resin found associated with Baltic amber, is pale yellow, transparent, and when first uncovered appears to be coated with a white powder. It is known to miners as "brittle amber" because it splinters easily. Because of its brittleness and its hardness of only 1 to 1.5, it has limited use in jewelry. The fracture is conchoidal, with a glassy luster, and a very resinous luster when polished. The specific gravity is 1.06 to 1.066, with a density of 1 to 2. When heated to 140 to 180 degrees centigrade, it softens and inflates, and, with continued heating, it melts. Gedanite is soluble in linseed oil and alcohol. It contains much less succinic acid than succinite. A chemical analysis of gedanite yields: carbon, 81.01 percent; hydrogen, 11.41 percent; oxygen, 7.33 percent; and sulfur, 0.25 percent. Enclosures are rare, but botanical debris, as small fragments of a pine-like wood, decomposed leaves and a few insects, have been found. Gedanite is thought to be the fossil resin of an extinct pine species, the *Pinites stroboides,* which resembles a five-needle white pine species.[66] The name gedanite comes from the Latin name for Danzig, *Gedania.*

Beckerite, named after one of the members of the firm of Stantien and Becker, who developed the amber industry, is another amber-like fossil resin found in the Baltic area. It is dark brown, soft, dense, non-fusible and usually found in lumpy, opaque or cloudy masses, only the edges of which are slightly transparent. Specimens remain opaque even after grinding and polishing. Beckerite is tough and it is difficult to pulverize, but when crushed, it produces a gray-brown powder. It is denser than Baltic succinite and contains only traces of succinic acid. It is thought to be the fossil resin of a leguminous tree. Among the miners, Beckerite is called "Braunharz" or "brown resin."

Stantienite, named after Stantien of the Stantien and Becker amber firm, is found mainly in the mining of the blue earth near Yantarnyy (Palmnicken). It resembles beckerite in its dark color and complete opacity, and presents a dull, dark black fracture surface. Specimens become strongly glittering after polishing. It is extremely brittle and more easily powdered than beckerite, producing a cinnamon-brown powder with less succinic acid than beckerite. Miners call it "Schwarharz," or "black resin," after its color.

Glessite is a light brown, almost opaque fossil resin, breaking with a conchoidal fracture displaying a greasy luster. It is 2 in hardness and lacks fluorescence. No plant or animal inclusions are found, but the presence of alpha-amyrin found by Judith Frondel in this resin suggests an angiosperm source.[67] Its name is derived from the ancient name for amber, *gles* or *glez.*

Krantzite is a soft fossil resin found in the coal areas of Saxony, as well as in the Baltic Kaliningrad area.

Schraufite is a reddish resin with the generalized chemical formula of $C_{11}H_{16}O_{12}$. It is found usually associated with lignite and jet in the Carpathian area of the Soviet Union and in certain sandstones of Austria.

Delatynite, found in areas of the Baltic and Carpathian regions, is a fossil resin mentioned in Soviet scientific literature without further details as to its appearance, size and properties.

REFERENCES

1. Haddow, J. G., *Amber, All About It.* Liverpool, England: Cope's Smoke Room Booklets, No. 7, 1891, p. 8.
2. Samples, C. C., "Amber," *England Mining Journal* 80: 250, 1905.
3. Sanderson, Milton R. and Farr, Thomas H., "Amber with Insect and Plant Inclusions from the Dominican Republic," *Science* 131: 1313, 1960.
4. *Ibid.*
5. Brouwer, P. A., *Economica Dominicano* 11, *16,* 1959
6. Hueber, F., Smithsonian Institute, personal communication after his return from Dominican Republic. The fossils were dated by Dr. Martin Buzas and Dr. Richard Cifelli and found to be lower Miocene or younger.
7. Sinkankas, John, *Gemstones of North America.* New York: Van Nostrand Reinhold, 1959, p. 675
8. Sanderson and Farr, "Amber with Insect and Plant Inclusions from the Dominican Republic," *Science* 131: 1314.
9. Bauer, Max, *Precious Stones* II: *55.* London: Charles Griffin and Co., 1904 (reprinted by Dover Publications, 1968).
10. Laufer, Berthold, "Historical Jottings on Amber in Asia," *Memoirs of the American Anthropological Association* I, *3:* 218, Lancaster, Pennsylvania, 1907.
11. *Ibid.,* I, *3:* 217.
12. Stevens, Bob C., *The Collector's Book of Snuff Bottles.* New York: Weatherhill, 1976, p. 185.
13. Chhibber, H. L., *Mineral Resources of Burma.* London: Macmillan & Co., 1934, pp. 90–91.
14. Williamson, George C., *The Book of Amber.* London: Ernest Benn, 1932, p. 116.
15. Stevens, *Snuff Bottles,* p. 185.
16. Williamson, *The Book of Amber,* pp. 219–220.
17. Langenheim, Jean, "Amber: A Botanical Inquiry," *Science* 163, *3:* 1160, 1969.
18. Bauer, *Precious Stones,* p. 555.
19. Chhibber, *Mineral Resources of Burma,* p. 85.
20. Bauer, *Precious Stones,* p. 55.
21. Chhibber, *Mineral Resources of Burma,* p.85.
22. *Ibid.,* p. 86.
23. *Ibid.,* p. 89.
24. *Ibid.,* p. 5.
25. *Ibid.,* p. 92.
26. Hunger, Rosa, *Magic of Amber.* London: N.A.G. Press, 1977, p. 83.
27. Cammann, S., *Substance and Symbol in Chinese Toggles.* Philadelphia: University of Pennsylvania Press, 1962, p. 256.
28. Sinkankas, John, *Van Nostrand's Standard Catalog of Gems.* New York: Van Nostrand Reinhold, 1968, p. 54.
29. Helm, Otto, *Mittheilungen über Bernstein: 14 über Rumanit, ein in Rumanien vorkommendes fossiles Harz.* Danzig: Schriften der Naturforschenden Gesellschaft, N. F. 7, 1891, p. 189.

30. Murgoci, George, *Gisements Du Succin De Roumanie avec un Aperçu Sur Les Résines-Fosiles (Succinite, Rumanite, Schraufite, Simetite, Burmite)*. Bucarest: L'Imprimèrie De L'État, 1903.
31. Liddicoat, R. T., Jr., *Handbook of Gem Identification*. Los Angeles: Gemological Institute of America, 1969, p. 395.
32. Williamson, *The Book of Amber,* p. 215.
33. Langenheim, "Amber: A Botanical Inquiry," *Science* **163,** *3:* 1157.
34. Strong, D. E., *Catalogue of the Carved Amber in the Department of Greek and Roman Antiquities*. London: The Trustees of the British Museum, 1966, p. 2.
35. Bauer, *Precious Stones,* pp. 535–557.
36. Webster, R., *Gems, Their Sources, Description and Identification*. Hamden, Connecticut: Archon Books, 1975, p. 513.
37. Savkevich, S. S., "State of Investigation and Prospects for Amber in U.S.S.R.," *International Geology Review* **17,** *8:* 921, Washington, D.C., 1975.
38. Buffum, Arnold, *The Tears of the Heliades or Amber as a Gem*. London: Sampson Low, Marston and Co., 1897, p. xvii.
39. *Ibid.,* p. xix.
40. *Ibid.*
41. Webster, Robert, *Practical Gemology*. London: N.A.G. Press, 1976, pp. 177–183.
42. Langenheim, "Amber: A Botanical Inquiry," *Science* **163,** *3:* 1161.
43. Williamson, *The Book of Amber,* p. 209.
44. Buddhue, J. D., "Mexican Amber," *Rocks and Minerals* **10,***170:* 171, 1935.
45. Sanderson and Farr, "Amber with Insect and Plant Inclusions from the Dominican Republic, *Science* **131:** 1313.
46. Ruzic, Raiko H., "Amber of Chiapas, Mexico, Parts 1 and 2," *Lapidary Journal* **27:** 1304 –1306 and 1400 –1406, 1973.
47. McAlpine, J. F. and Martin, J. E., "Canadian Amber—A Paleontological Treasure Chest," *The Canadian Entomologist* **101:** 819, August 1969.
48. Langenheim, "Amber: A Botanical Inquiry," *Science* **163,** *3:*1157–1169.
49. McAlpine and Martin, "Canadian Amber," *The Canadian Entomologist* **101:**819.
50. Carpenter, F. M., "Fossil Insects in Canadian Amber," *University of Toronto Studies, Geological Series, No. 38:* 69, 1935.
51. Carpenter, F. M., Flosom, J. W., Essig, E. O., Kinsey, A. L., Brues, C. T., Boesel, M. W., and Ewing, H. E., "Insects and Arachnids from Canadian Amber," *University of Toronto Studies, Geological Series, No. 40:* 7–62, 1937.
52. McAlpine and Martin, "Canadian Amber," *The Canadian Entomologist* **101:** 819.
53. Walker, T. L., "Chemawinite or Canadian Amber," *University of Toronto Studies, Geological Series, No. 36:* 5–10, 1934.
54. Sinkankas, *Gemstones of North America,* p. 601.
55. Saunders, W. B., Mapes, R. H., Carpenter, F. M., and Elsik, W. C., "Fossiliferous Amber from the Eocene (Claiborne) of the Gulf Coastal Plain,"*Geological Society of America Bulletin* **85:** 979, June 1974.
56. *Ibid.,* **85:** 979–980.
57. King, Albert A., "Texas Gemstones," *Report of Investigations, No. 42:* 18, Texas State Geology Department, 1961.
58. Jewell, W. B., *Geology and Mineral Resources of Hardin County, Tenn.,*Tennessee Division of Geology, Bulletin 37, pp. 94–95, 1931.
59. Cockerell, T. D., "Some American Fossil Insects," *Proceedings, U. S. National Museum* **51:** 89, 1916.
60. Taggart, Joseph E., New Mexico Bureau of Mines and Minerals Resources, personal communication.
61. Goldsmith, E., *Proceedings, Academy of Natural Science*. Philadelphia, 1879, p. 40
62. Sinkankas, *Gemstones of North America,* p. 599.
63. Langenheim, "Amber: A Botanical Inquiry," *Science* **163,** *3:* 1157.

64. Sinkankas, *Gemstones of North America,* p. 680.
65. MacFall, Russell P., *Gem Hunter's Guide.* New York: Thomas Y. Crowell, 1975.
66. Zalewska, Z., *Amber in Poland.* Warsaw: Wydawnictwa Geologiczne, 1974, p. 33.
67. Frondel, Judith W., "Amber Facts and Fancies," *Economic Botany* **22,** *4:* 381–382, October–December 1968.

ADDITIONAL REFERENCES

Allen, Jamey, "Amber and Its Substitutes, Part II, Mineral Analysis," *The Bead Journal* **2,** *4:* 11–12, Spring 1976.

Arem, Joel, *Gems and Jewelry.* New York: Bantam Books, 1975, pp. 91–93.

Harrington, B. J., "On the So-called Amber of Cedar Lake, N. Saskatchewan, Canada," *American Journal of Science* **42,** *3:* 332–338, 1891.

Helm, Otto, "Mittheilungen Über Bernstein, 2, Glessit, 3 über sicilianischen und rumanischen Bernstein," *Schriften der Naturforschenden Gesellschaft zu Danzig.* N.F. 5, Danzig, 1881, pp. 291–296.

Helm, Otto, "On a New Fossil, Amber-like Resin occurring in Burma, from Upper Burma," *Rec. Geol. Surv. Ind.* **25:** 180–181, 1892.

Langenheim, Jean H., "Present Status of Botanical Studies of Ambers," *Botanical Museum Leaflets.* Cambridge: Harvard University Press, **20,** *8,* May 1964.

Noetling, F., "Preliminary Report on the Economic Resources of Amber and Jade Mines Area in Upper Burma," *Rec. Geol. Surv. Ind.* **25:** 130–135, 1892.

Zahl, Paul A., "Amber, Golden Window of the Past," *National Geographic* **152,** *3:* 423–435, September 1977.

8
Natural Resins Resembling Amber

The oldest known substitute for amber is *copal,* a resin which is very similar in appearance. Before synthetic resins were produced, copal was the principal substitute and, in earlier periods, was worked into ornaments in those cultures having ready access to supplies of the rough material. Ancient Egyptian graves commonly contained lumps of copal, providing evidence of the earliest known use of copal in jewelry for royalty. Articles made of semi-fossilized resin, including a double ring with the resin engraved in the form of the royal seal, were found in the tomb of King Tutankhamum. Two large scarabs—one carved on the surface with a relief design of a bird—and two necklaces—one with graduated resin beads and one alternating resin and lapis lazuli beads—were also found, as well as resin earrings, hair rings and a knob for a box. The resin in all of these objects is dark red, perhaps from oxidation, and very brittle. Tests showed the resin was readily soluble in alcohol and acetone, both of which scarcely affect true amber.[1]

The Chinese also used resins other than amber for carving and ornaments. The accompanying photograph shows an old Chinese toggle button made of copal containing insect inclusions. Since toggle buttons were not only used to fasten outer garments but were also decorative and often held symbolic meanings, the symbolism attached to the resins and inclusions attributed to the desirability of such buttons. As can be seen in Fig. 8–1, the copal objects, even after having been polished, develop a finely checked or crazed surface, often in just a few years, as a result of the escape of volatile substances.[2]

The word *copal* was derived from the Spanish word *copalli,* meaning "incense," and refers to a group of resins exuded from various tropical trees and used principally in varnishes. Such resins are soluble in oils and organic liquids, but are insoluble in water. In making varnish, copal is melted in a copper pan, and at 212 degrees farenheit, water contained within the resin evaporates, with a considerable amount of gases evolved as the material

219

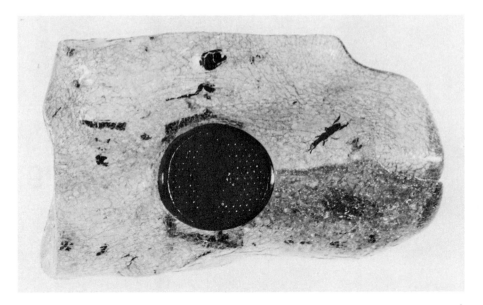

Fig. 8–1. Antique Chinese toggle button made of copal with several insect inclusions. Note the heavily crazed surface. Actual size: 2½ inches x 1¾ inches. (Author's collection.)

decomposes. The copal preserves its yellow color but becomes more soluble.[3]

During the time when natural varnish resins were in great demand, several regions of the world exported various types of resins (see Fig. 8–2). Raw copal resin was obtained by directly tapping the trees, and was collected for shipment to varnish manufacturers. However, synthetic resins have now replaced most natural resins, and only those varieties commonly confused with amber will be discussed further.

Copals occur both in a semi-fossilized form called "true copal" and in freshly obtained gum form called "raw copal." The semi-fossil kinds range from over 1000 years to as little as 100 years old. They are either transparent or translucent and typically are yellow or brown in color. Copals are softer than amber, but are very brittle and break in conchoidal fractures. When heated, they emit a distinctive resinous odor and burn with a smoky flame. The refractive index (1.54) and specific gravity (1.06 to 1.08) are similar to those of amber, but the lower hardness and easy solubility can be used to identify copal when it is masquerading as amber. A drop of ether on copal causes it to become sticky, whereas on true amber no reaction is noted. Extensive crazing also suggests that the material is copal rather than true amber. Copals fluoresce white under short-wave ultraviolet light.[4,5]

Insects, leaves and other botanical debris are found in both semi-fossil copals and recent copals. These enclosures are always of recent, currently living varieties rather than the primitive, extinct forms found in amber. Enclosures may be artificially placed in either copal or amber by inserting the insect between two layers of the substance and fusing them together. The

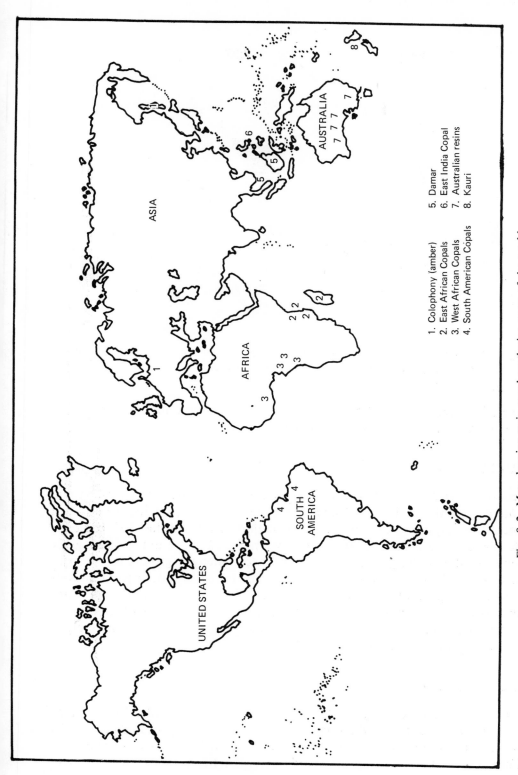

Fig. 8–2. Map showing main copal producing areas of the world.

1. Colophony (amber)
2. East African Copals
3. West African Copals
4. South American Copals
5. Damar
6. East India Copal
7. Australian resins
8. Kauri

Fig. 8-3. Lump of copal from Congo region containing several ants and a present day grasshopper *(Orthoptera)*. Actual size of grasshopper: 12mm.

layered material may be detected by careful examination under a microscope or with a high power jeweler's eyepiece.

ZANZIBAR COPAL

Semi-fossilized copal from Zanzibar, sometimes termed "jackass" copal (*chakasi*) by the natives, is found embedded about 2 to 4 feet (0.6 to 1.2 meters) underground over a wide area of East Africa on the coast opposite the island of Zanzibar, but the trees that produced the resin no longer grow there. Shallow-mining, during the rainy season when the ground is soft, produces pieces from pebble size to masses weighing several ounces. Occasionally, lumps up to 4 to 5 pounds (1.8 to 2.3 kilograms) may be found. Zanzibar raw copal is also collected from the living trees (*Trachylobium verrucasum*) or obtained near the roots and on the ground. It is commercially used for manufacturing varnish in China and India.[6,7]

KAURI COPAL

The New Zealand (see Fig. 8-6) Kauri copal was and is exuded by the Kauri pine, *Agathis australis,* which belongs to the family of Araucariaceae and flourishes in the northern region of New Zealand, the only area where the Kauri has grown for many thousands of years. Kauri trees often live over 1000 years and grow to a giant size, with a trunk that ranges from 5 to 12

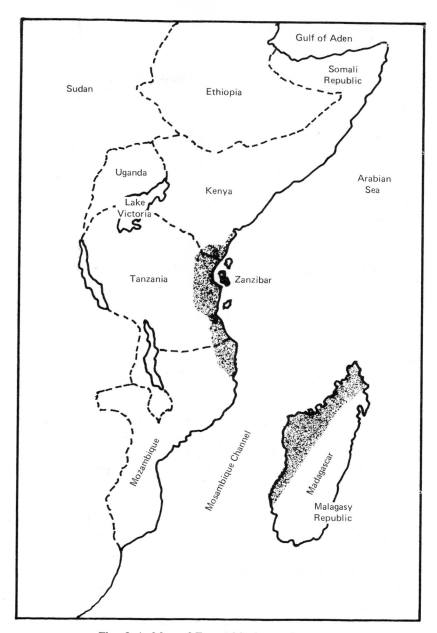

Fig. 8-4. Map of East Africa's copal areas.

feet (1.8 to 4 meters) in diameter and a height of 120 to 160 feet (40 to 50 meters).

In certain areas where a number of forests flourished, two or more layers of copal resin may be found in the ground. In some instances, Kauri copal has been found under a layer of sandstone and limestone as deep as 300 feet (100 meters) below the surface.[8] Kauri resin taken from the ground, though extremely old, is still soluble and therefore is considered to be in a semifossilized state. Its color ranges from yellow to reddish yellow to brown. It sometimes contains many insect inclusions.

In the 1850's, manufacturers of varnish began using Kauri resin to pro-

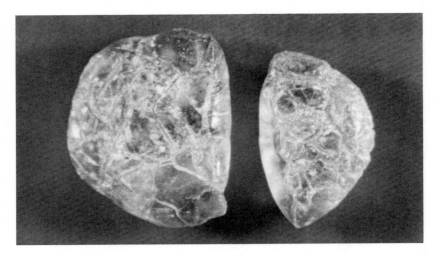

Fig. 8-5. Raw lumps of Kauri gum. *(Courtesy Kuhn's Baltic Amber Collection.)*

duce one of the better quality varnishes, with the industry reaching its peak about the turn of the century. Though many trees were tapped for the purpose of collecting the raw gum, much buried material was "mined" in a simple process. A long, pointed, steel rod was used to probe the earth, and was thrust into various spots until a lump of Kauri resin was felt, at which time spades were used to dig up the lump. In contrast, when prospecting in peat bogs or swampy areas, a hook was inserted into the soft mud and the resin lumps snagged to bring them to the surface. Exceptionally large and clear specimens were prized by the "gum diggers," who often polished such pieces in natural form or carved them into ornaments.

Kauri does not contain succinic acid. Its specific gravity is 1.05 and it melts between 360 and 450 degrees farenheit (187 to 232 degrees centigrade). As a result of this low melting point, the friction produced during polishing commonly causes the surface to become sticky. Also because of its softness, it does not polish well; yet, in 1912, Penrose[9] reported that clear, transparent varieties were used as substitutes for amber in mouthpieces for pipes, cigars and cigarette holders, and that a certain amount was carved into small ornaments.

OTHER COPALS

Raw copal from Sierra Leone is produced by a leguminous tree of the family Caesalpiniaceae, *Copaifera guibourthiana,* which is the same tree that produces the Congo copals. The resin is often found in large peat swamps, and, in the past, was cleaned by washing in a rotating drum. Following this treatment, the copal was sorted and sent to various varnish manufacturing factories.[10]

Congo copals, similar to those from Sierra Leone, are usually found in the hardened semi-fossilized form.

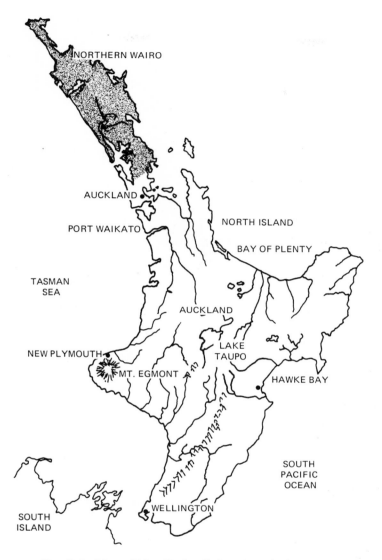

Fig. 8-6. Map of New Zealand's kauri-producing area.

Manila copals differ from the West African copals in being produced by trees of the genus *Agathis* in Indonesia and the Philippines, the principal regions where this copal is collected from ground deposits or from living trees intentionally tapped to produce a resin flow. Various names are used to denote the resin in different stages of hardness, with the hard, semi-fossilized forms being called "boea" (which, despite its hardness, is alcohol-soluble). The native Philippine name for resin somewhat hardened by delaying its collection for one to three months after tapping the tree is "loba." A still softer kind gathered about two weeks after trees are tapped is called "melengket." Another form of hard, semi-fossilized Manila copal is found in Borneo and is called "pontianak."

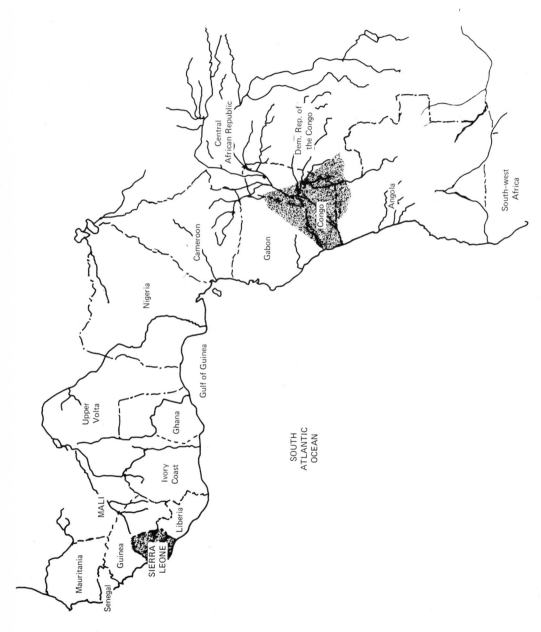

Fig. 8-7. Map of West Africa's copal-producing areas.

The tropical tree, *Hymenaea courbaril,* as well as other species, produces copal resin in Colombia and Brazil. The resin of *Hymenaea c. is* typically greenish yellow in hue.[11]

DAMMAR RESIN

Dammar, a clear, pale yellow copal-like resin, which is sometimes called "copal," melts at about 140 degrees centigrade and is soluble in aromatic hydrocarbons (such as benzene) but is insoluble in alcohol. It is one of the harder resins used in making lacquers and clear varnishes and is chiefly obtained from dipterocarpaceous trees of southern Asia, especially in Malaya and Sumatra. The resin may be either semi-fossilized or recent.

CURRENT INFLUX OF "AFRICAN AMBER"

At present, a lively interest in old African art and ornament has resulted in the appearance of much "African amber" and many African glass trading beads. "African amber" or "copal-amber" are popular names attached to oblate or barrel-shaped beads supposedly made of copal resin found along the east coast of Africa and elsewhere on the continent, and used to some extent by native tribes for jewelry. Some such beads were made of true amber from the Baltic and imported during the period of slave-trading on the coast of East Africa. Erikson[12] states: "Beads, of great size, brown, yellow, blue and colored, are reported as trade items for African gold, ivory and slaves during the 16th century." African chiefs during this period adorned themselves, their many wives and their children, in great quantities of beads. The slaves in Zanzibar were bedecked with beads to enhance their appearance and to attract buyers. This trade continued until 1899, when the last recorded slave ship was wrecked off Wasin Island, not far from Mombasa, East Africa.

In this connection, the *Dispatch,* a ship leaving Bristol, England on September 30, 1725, records a variety of unusual beads in its hold for the purpose of trading, such as "1378 lbs bugles [Venetian beads]; a sort of bugle called Pagant; Maccatone, that is, beads of two sorts; Christal pipe beads; and yellow amber." Today, we cannot be sure of the meaning of all the names for the beads listed, but we do know that amber was included in this early trade with Africa.[13]

Most amber sent to Africa was opaque, yellow, oblate-shaped large pieces considered by Europeans to be of an inferior type. Amber beads were strung along with colorful glass beads by the natives. Sometimes only one large amber bead was placed in the center of the necklace. Occasionally, a strand of "African amber" currently on the market will contain a few of these true amber beads.

Since about 1971, long, heavy strands of amber-like beads appeared on the United States market, imported from Africa. These beads are generally cloudy or opaque, ranging from yellow to brown in color. Some have swirls of transparent material within the otherwise cloudy mass. Shortly after the first influx of large beads, a smaller, barrel-shaped bead was introduced, very even in size and color, as compared to earlier strands, which were quite diverse in respect to the color and quality of the beads on a single strand.[14]

In local bead shops, the strands are romantically represented as "African amber" or copal from the depths of the jungle, used by African natives as ornaments in the past. Fascinating tales are told of primitive methods of digging for "amber" and of savage barter as members of tribes attempt to obtain their "African amber." Yet, when tested, such beads generally prove to be neither amber nor copal. One can only assume that because of increased demand for "African amber," synthetics are being used to produce "pseudo-copal," or an imitation of a substitute! (See Color Plate 51.)

The hot point test, which is accomplished by touching a hot needle or wire to an inconspicuous spot near the perforation of a bead, aids in identifying plastic imitations. An unpleasant acrid odor will be emitted from synthetic materials, rather than the resinous scent from either copal or amber. If a fine thread of the material adheres to the hot point as it is pulled away, the substance is most likely a thermoplastic. Such beads are often lighter than amber and will float in a solution of salt water.

Around 1974 to 1975, "African amber" beads in a pale yellow shade, as well as a turbid red, began appearing on the market and were fashioned in a variety of unusual shapes. The origins of these beads were given as several African countries, including Mali, Morocco, Nigeria and Ethiopia, which are not sources of genuine copal. When tested, these beads usually were found to be either thermoplastics or thermosetting plastics.

The dark reddish "African amber" beads referred to as heat-reddened "copal," again, more often than not, prove to be imitations, mostly made from thermosetting plastic. Some cloudy yellow thermosetting plastics turn to red when heated in an oven at a temperature of 350 degrees farenheit for 10 to 20 minutes, the color penetrating completely throughout the bead.[15] If similar treatment were given to true copal, the copal would soften and begin to decompose, giving off a resinous vapor.

Erikson, in her work, *The Universal Bead,* stresses the possibility of finding plastic reproductions of older bead forms: "In the markets of faraway Pakistan, plastic beads are combined with brass or gilt reproductions of older bead forms and long strands of plastic amber beads may be seen hanging in the Arabian market stalls in Acre, Israel."[16]

Originating in the Middle East and Afghanistan, small barrel-shaped beads, called "prayer beads," have been imported recently and distributed widely. Most are orange to orange-brown in color, and semi-translucent, with a grainy appearance under magnification. They are usually uniform in appearance and shape. Occasionally, flat edges of the beads are stained a darker color on the surface. Small orange beads of this material are marketed as either "Egyptian" or "Afghanistan amber" rather than copal; once again, the majority prove to be either thermoplastic or thermosetting plastic imitations—not amber. Sometimes they are described as "amber

Fig. 8–8. "Afghanistan amber" and "prayer" beads made of synthetic material to imitate copal. Note the uniform appearance of all the beads.

from Russia, cut into beads in Afghanistan."[17] Such barrel-shaped beads are commonly seen in conjunction with old Middle East silver beads.

Prayer beads are often strung with an unusual central bead designed to be used in forming a tassel which hangs from the center of the strand. Authentic Moslem prayer beads of the past were required to be made of amber and contained 99 beads, one for each martyr of the Moslem faith.

Jamey Allen, historian of ornamentation and authority of "African amber," relates an interesting experience associated with such beads:

In East Africa, in market places and elsewhere, were many strands of amber-like beads. They had a very even, manufactured look to them, were barrel shaped, yellow or brown in color, and were called "Somali amber." After making a few inquiries, [a friend] was informed by the Nairobi Museum in Kenya that the natives cut these beads from long, broom-handle-sized rods. These rods have been imported from Europe and are made of synthetic resin, or plastic. They are represented to the natives as amber, and not knowing what amber truly is, or what the composition of the rods is, the natives repeat the story. My friend brought back a strand of Somali amber, as well as a strand of "real amber," which is probably from Zanzibar, and may be authentic copal. The plastic "Somali amber" strongly resembles most African and Middle Eastern "copal" in appearance.[18]

Allen advises that when examining copal-type beads, a fairly uniform grain suggests the beads were formed or pressed, because "a uniform grain running parallel to the axis of many similar, large-sized beads indicates that they were originally one long piece."[19]

REFERENCES

1. Lucas, A., *Ancient Egyptian Materials and Industries*. London: Ed. Arnold and Co., 1934, p. 337.
2. Cammann, S., *Substance and Symbols in Chinese Toggles*. Philadelphia: University of Pennsylvania Press, 1962, pp. 78–79.
3. Rebaux, M., ["Natural Amber"], *Annales de Chimie et de Physique*, Paris, 1880.
4. Anderson, B. W., *Gem Testing*. New York: Emerson Books, 1948, p. 211.
5. Webster, Robert, *Gems, Their Sources, Descriptions, and Identification* I: 514. Hamden, Connecticut: Archon Books, 1975.
6. Reed, A. H. and Collins, Tudor W., *New Zealand's Forest King—The Kauri*. Wellington, Auckland: Consolidated Press Holdings, 1967.
7. Hornaday, W. D., "Kauri Gum Deposits of New Zealand," *Mining Press* 110: 181–182, January 1915.
8. Reed and Collins, *The Kauri*.
9. Penrose, R. A. F., Jr., "Kauri Gum Mining in New Zealand," *Journal of Geology* 20, *1:* 43–44, January–February 1912.
10. Barry, T. Hadley, *Natural Varnish Resins*. London: Ernest Benn, 1932, p. 51.
11. "Copal," *Encyclopedia Americana, International Edition* VI: 458. New York, 1975.
12. Erikson, Joan Mowat, *The Universal Bead*. New York: W.W. Norton and Co., 1969, p. 58.
13. *Ibid.,* p. 60.
14. Allen, Jamey, "Amber and Its Substitutes, Part 3: Is It Real? Testing Amber," *The Bead Journal* 3, *1:* 20, Winter 1976.
15. *Ibid.,* 3, *1:* 26.
16. Erikson, *The Universal Bead,* p. 122.
17. Allen, "Amber and Its Substitutes, Part 3: Is It Real? Testing Amber," *The Bead Journal* 3, *1:* 26.
18. *Ibid.,* 3, *1:* 22.
19. *Ibid.*

Part 4
Commercial Aspects

9
Imitations of Amber and Their Discrimination

Without question, there are far more substitute and make-believe gemstones in circulation throughout the world than there are natural, properly identified stones. This is an ancient state of affairs persisting since the beginning of man's experience with gems. Pliny's *Natural History* summed up the situation beautifully for all time. "Truthfully," he wrote, "there is no fraud or deceit in the world which produces larger gain and profit than that of counterfeiting gems." Very likely there are more profitable endeavors and very likely the statement is a little strong and uncharitable toward those engaged in this trade. Most frequently gem counterfeiting is not intended as fraud. When fraud occurs it is not the fault of the materials or their producers necessarily, but rather the motives of the buyer or seller. Gem counterfeiting caters to a very large market composed of those thousands of people who feel they cannot afford the price of an attractive natural gem. Also, there are many who appreciate the decorative value of gems, but prefer to make their investments elsewhere.

Paul Desautels,[1]
Smithsonian Institute

THE NATURE OF PLASTICS

Since ancient times, attempts have been made to imitate amber. Ancient Chinese literature describes an imitation amber made by boiling chicken eggs with "the fat of the dark fish" and suggests using amber's electrical properties to distinguish this imitation from the genuine material. As time passed, imitations of amber were continually improved, and today synthetic products have many characteristics similar to those of genuine amber, including external appearance and even some electrical properties.

Amber and other natural resins are composed of carbon, hydrogen, oxygen and other elements which are the same as those used in the manufacture of plastics or synthetic organic compounds of high molecular weight (*polymers*). The word "polymer" is of Greek origin, meaning "many parts." Polymers are built up from *monomer* units, or one-part molecules, in the process known as polymerization, where such molecules are combined to form long chains. Though plastics or polymers are produced in hundreds of varieties, there are basically two groups: the *thermoplastics* and the *thermosetting plastics*.

Thermoplastics require heat to soften and mold them, and they can generally be reheated and remolded if necessary. No chemical changes occur during the essentially physical process of softening and solidifying. In contrast, *thermosetting plastics* harden or "cure" when heat is applied, and they generally cannot be remolded by reapplication of heat, since such irreversible hardening is brought about by a chemical reaction and is not merely a physical process. Both types have been used as imitations for amber.

CELLULOIDS

The first synthetic plastic widely used as an amber imitation was *celluloid,* or cellulose nitrate, invented in 1867 by John Westly Hyatt, and originally made to imitate ivory. The high cost of the natural substance after the Civil

Fig. 9–1. Celluloid imitation of tumbled amber necklace.

Fig. 9–2. Thermoplastic imitation of antique amber beads. When touched with a hot point, the bead melts easily and a fine thread of the plastic adheres to the point.

War led manufacturers to search for a substitute. A $10,000 prize was offered to anyone who could invent an appropriate imitation. Hyatt, an inventor from Albany, New York, began experimenting with cotton and acids, and, by 1868, he had produced a good imitation especially suitable for billiard balls. He called his invention celluloid, and this was the first plastic made in the United States. Celluloid quickly became an important material for making many other articles, including handles, combs, buttons, buckles, false teeth and photographic film. Imitation amber beads made of celluloid were called ''ambre antique.''[2,3,4]

Celluloid is thermoplastic and thus is repeatedly fusible, softening when heated and hardening when cooled, but decomposing at high temperatures. It is made by treating cellulose, usually cotton, with nitric acid and sulfuric acid, with camphor added to make the product less brittle and easier to mold.

''Ambre antique,'' or celluloid imitations of amber, can be detected by examining the swirl lines commonly appearing as sharp delineations between clear and cloudy areas. Mold lines sometimes may be detected. Parings can be shaved from celluloid with a knife, but genuine amber splinters or crumbles when cut.[5]

Fig. 9-3. Currently, light thermoplastic beads are on the market, represented as melted amber. These beads float in plain tap water and the plastic resin adheres to the hot point.

When touched with a hot point, celluloid adheres to the point and may flare into a bright flame with emission of camphor odor. Because genuine amber has a higher melting point, it does not adhere to the hot point and burns only slowly, giving off a "piney" odor. Celluloid is readily soluble in sulfuric ether, though amber can lie in this liquid for 15 minutes without serious damage.

Because of the dangerous inflammability of celluloid, it could not be used in smoker's articles. Manufacturers attempted to eliminate this major disadvantage and finally devised a "safe" celluloid, made by treating cellulose with a mixture of glacial acetic acid and sulfuric acid. This produced cellulose acetate, which is still extensively produced and used under a variety of trade names.

Though celluloid is not used today as an amber imitation, older "amber" beads dating from the late 1800's should be tested for specific gravity to determine whether the beads are celluloid or genuine amber. A saturated solution of salt water with a density of about 1.13 floats amber, but the heavier celluloid (density about 1.38) sinks in this solution.[6] The refractive index of celluloid is 1.50, compared to amber at 1.54, but the substitute may be just as easily identified by placing a drop of 5 percent solution of diphenylamine in sulfuric acid on the surface of the specimen: if the material is celluloid, a blue color is produced; if it is amber, no color change takes place.[7] Under ultraviolet light, celluloid fluoresces yellowish white.

BAKELITE

In 1909, a Belgian-American chemist, Leo Hendrik Baekeland, while searching for a substitute for shellac, invented the first and most versatile of all thermosetting plastics, a pheno-formaldehyde condensation resin which was given the name *bakelite*. Baekeland produced a dark syrup which hardened when heated, could be molded into any shape, was not dissolved by common solvents, was non-flammable and was inexpensive. It soon became the most widely used synthetic resin for imitating amber.[8,9]

The general process for producing thermosetting plastics is by reacting phenol (or its derivatives) with formaldehyde in a condensation reaction, through three stages, each producing a product with different physical and chemical properties. The first stage, polycondensation, is carried out in water and leads to water-soluble or water-dispersible intermediates. The second stage obtains a resinous, insoluble and difficult to fuse mass which can be converted into fine, dry powder; this is sold to manufacturers of synthetic amber and molded and shaped into desired forms. Application of heat and pressure in the third stage converts the product to a resin of maximum hardness and permanent infusibility. If this material is burned, it does not soften or melt, nor does it burn by itself, but simply carbonizes into a black cindery material. Thermosetting plastics are generally dense—much more so than genuine amber—with their specific gravity ranging from about 1.11 to as much as 1.55 when fillers or pigments are added.

Fig. 9-4. Estate beads purchased as true amber, but tested to be heavy thermosetting plastic (with an acrid odor when touched with a hot point).

It must be remembered that synthetic resins have been in use for over a hundred years, and it is important, when examining antique beads, to test for their presence. In 1920, when amber reached a new high in popularity, long imitation "amber" necklaces were produced in a variety of faceted bead shapes and colors, the most popular being a red imitation of highly prized Sicilian amber. Deep black and dark ruby-red beads were also produced to resemble rumanite amber. The majority of deep red "amber" beads found in old estates dating from the early 1900's are made of bakelite, and their external appearance may closely resemble the genuine article, making it very easy to mistake an imitation for true amber. (See Color Plate 52.)

Another useful clue to imitations depends on the fact that amber is softer than most plastics. Therefore, amber, with constant rubbing of the stringing material, in time wears away around the hole in the bead. Sometimes the edges chip as the beads rub together. Since bakelite is tougher, even beads made of bakelite that have been strung for 60 or 70 years or more still have clean smooth holes. Furthermore, old amber beads commonly develop shallow crazing over the surface, resulting from temperature changes over the years, whereas plastic beads still retain a smooth polish.

Other physical properties of phenolic resins can be used to identify them. Although the different types of plastics also differ in their specific gravities,

Fig. 9–5. Red bakelite plastic faceted beads of this type are extremely common, dating from the early 1920's. Beads retain their sharp facets and show little wear at the edges of the perforations. The material is transparent, with no crazing.

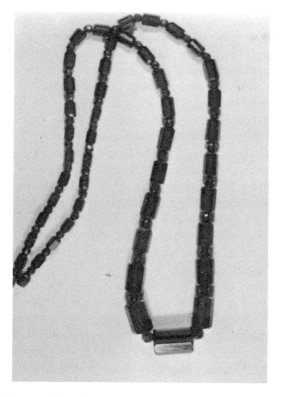

Fig. 9–6. Red bakelite plastic beads were faceted in a variety of shapes and were strung into long necklaces.

the denser types most commonly used for imitating amber average 1.26 in specific gravity, much above that of amber at 1.05 to 1.09. On the other hand, the newer thermoplastics are generally lighter than amber, and the simple flotation test in saturated salt solution will not separate them from amber. As an added distinction, the refractive index of 1.60 to 1.66 for bakelite is well above amber at 1.54, but testing carved pieces or mounted specimens without flat or easily accessible surfaces creates difficulties in determining either refractive index or specific gravity. If these tests are impossible, the hot point test may be used on some inconspicuous spot to detect the acrid odor and resistence to burning of bakelite.

Another useful test for bakelite involves boiling a scraping or peel in about 2 milliliters of distilled water to dissolve a little phenol; adding to the solution a pinch of 2.6 dibromoquinonechlorimide;[10] and dipping a glass rod into a solution of caustic soda and, retaining a very small drop on the rod, using it to stir the water. If the water turns blue, the presence of phenol is indicated and the material is a phenolic resin.[11,12,13]

CASEINS

Casein, a hardened milk protein, was occasionally used as an amber imitation. It was first made, in 1890, by A. Spitteler and W. Krische, of Hamburg, Germany, by condensing milk protein with formaldehyde to produce an insoluble, tough material. By 1914, England, France and Germany were making the plastic under the name of *galalith* (milkstone) or *erinoid*. Because of its non-flammability, casein promised to be a substitute for the dangerously flammable cellulose nitrate. It could also be produced in lighter shades of color than phenolic resins. In 1919, it was introduced into the United States, but found to be unsuitable for most uses other than making buttons. Cheap, rather crude beads imitating amber were made by adding fillers to the plastic.[14]

Caseins are denser than amber, with a specific gravity of 1.33, and the refractive index is 1.55. Under ultraviolet light, the material fluoresces white. A drop of nitric acid on the surface results in a bright spot, but because the body color of the imitation is often yellow, the change in color may be difficult to see. Caseins are sectile but tough. A touch of a hot point to casein produces a typical burned milk aroma.[15] No frictional electricity is developed upon caseins.[16]

POLYSTYRENES

Polystyrene, a thermoplastic, is gaining popularity for costume jewelry because of its suitability for injection molding. It is less dense than amber (specific gravity 1.05) and has a refractive index of 1.50. Its low specific gravity poses some problems in identification when using the usual flotation test. Jamey Allen suggests a refinement that was found to be useful in separating imitation substances from genuine amber, as follows. Fill three ten-ounce glasses with water. One tablespoon of salt is added to the first glass; two tablespoons to the second; and three tablespoons to the third. The lighter thermoplastics float in all three salt solutions, and some may even float in water without salt. Amber generally floats in only the second or third solutions, depending on the specimen, while bakelite and the heavier synthetic resins sink in even the most saturated salt solution.[17]

In addition to flotation, another reliable test for identification of polystyrene is its ready solubility in ketones, benzene, toluene and other aromatic hydrocarbon liquids.

BERNIT (BERNAT)

Gebhardt Wilhelm, a jewelry manufacturer in Germany, produces imitation amber jewelry called ''Bernit'' or ''Bernat.'' This material closely resembles amber because it contains ''stress spangle'' inclusions like those in genuine amber, and proper coloration. It is also similar in refractive index to amber, but is heavier, with a density of about 1.23. The discoidal stress spangles mentioned are circular, but lack the radiating lines within each fissure typical of those found in genuine amber. Under magnification, the circular fissures appear too clear, as they are much more transparent than those found in genuine amber. In addition, some of the fissures in the imitation are unnaturally curved or bent, unlike those in genuine amber, which are almost invariably flat. When a Bernit imitation is compared to a lump of genuine amber possessing the same sort of spangling, the difference between true amber and this imposter can be detected. In spite of these differences, Bernit imitations are extremely realistic and must be examined carefully to identify them.

Bernit is also available in pieces containing plant fragments and insects. Although these pieces closely simulate natural amber with inclusions, the imitation often can be detected because the plastic surrounding the insect is too clear and gives no signs of the insect having been embedded alive. In true amber, small air bubbles emitted from the insect's respiratory system are found close by; in some cases, the insect's struggles may have formed swirls in the resin. Large insects are particularly suspect and should be examined with care.[18]

POLYBERN

An amber imitation made by combining polyester resin and small pieces of real amber is called "polybern," and was developed recently in Germany. It is produced in Poland and Lithuania for making sculptures, souvenirs and

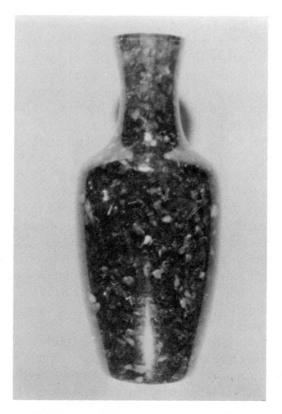

Fig. 9-7. Vase made of polybern, amber chips set in plastic.

Fig. 9–8. Necklace of polybern beads from Poland. (H. Even's collection.)

inexpensive jewelry. It is easily confused with genuine amber because it does contain embedded natural amber chips. The name "polybern" was derived by combining the words *polyester* and *Bernstein* (the German word for amber), the two substances of which the material is composed. "Polybern" is manufactured into oval beads for necklaces, long oval cabochon shapes for bracelets and a variety of rectangular shapes for brooches.

A similar imitation, but containing less genuine amber within the synthetic matrix, is currently flooding the market. This material is manufactured in Poland, and the bulky, cubic beads made from it are formed in graduated molds, the molds being filled half-full with a synthetic resin, followed by a layer of small amber chips, and finally filled with additional synthetic resin. The resulting cubes are perforated and strung in graduated sizes, with the center bead often as large as 1 by 1 by ½ inch (25 by 25 by 12.5 millimeters). Careful examination of the larger beads readily reveals the layered structure.

Another imitation amber product manufactured in West Germany and recently introduced into the market combines polyester resin with amber dust. It is attractive, closely resembles amber, and is practically impossible to distinguish from the natural material without careful examination and testing. When touched with a hot point, the polyester-amber dust product even emits the piney odor of true amber because of the abundance of amber dust, and, depending on the matrix resin used, it may float in a saturated solution of salt water.

West German laboratories are becoming so skilled in producing amber substitutes that physically and optically closely resemble real amber that only by using sophisticated tests and instrumentation beyond the resources of the ordinary gemological laboratory can the imitations be identified as such. Therefore, in 1968, courts in West Germany established consumer protection laws requiring that customers be given the facts regarding the type of stone purchased and the tests to establish its identity. Furthermore, such ar-

Fig. 9–9. Close-up of polybern bead to show the appearance of the amber chips within the synthetic resin.

ticles must be marked as natural or synthetic. When one purchases "amber" articles from West German manufacturers, one should look for such labels.

SLOCUM IMITATION AMBER

A United States firm, J. L. Slocum Laboratories of Royal Oak, Michigan, recently introduced an imitation amber material which lapidaries can purchase in block form. It is a transparent synthetic resin produced in both pale orange and red shades and is available with either insect inclusions or with discoidal sun-spangle fissures. The insect inclusions are, of course, of recent origin and are much more numerous than would be found in either Baltic or Dominican natural resins. The circular fissures can be distinguished in most specimens from those in natural amber because of their frosted appearance.

The Gemological Institute of America (GIA) reports the refractive index of the Slocum imitation to be 1.50 to 1.55, or very similar to that of natural amber. It has a specific gravity of 1.17, heavier than true amber, and, with a hardness of 3, it is harder than amber. When touched with a hot point, it emits the odor of burned fruit, in contrast to the piney odor of amber, indicating that it may be made from an acrylic, styrene or polyester resin. Chemical analysis of the resin is not available. (See Color Plate 53 for Slocum imitation.)

SUN-SPANGLE IMITATIONS

Many manufacturers of imitation amber are attempting to duplicate the popular "sun-spangled" amber. How the "sun-spangle" inclusions in imitation amber are created has not been divulged; one manufacturer simply states that he "cross-linked the molecules in a thermosetting plastic." With such little information regarding the method used, a sample of the plastic imitation was shown to a physicist conducting research using the laser and a high-speed photoflash (1000 joule 1 millisecond flash). He stated that the "sun-spangle" markings resemble the stress fissures that often occur accidentally when a lucite rod is placed inside the spiral of the flash lamp and the lamp excited (flashed). Minute explosions tend to occur within the mass of the lucite, and because of the hardened structure of the material, direction is given to the resulting fractures, causing them to appear as discoidal inclusions. Of course, other, less expensive, methods to produce stress fissures within polymers are more likely to be used by manufacturers of imitation amber.

GENERAL RECOMMENDATIONS FOR DETECTING IMITATIONS

When purchasing an "amber" article where it is infeasible to perform the more reliable tests of refractive index and specific gravity, careful examination of each bead can often detect imitations. Modern synthetics are so similar to the natural material that all articles sold as amber should be compared with a known piece of genuine amber. Carrying a piece of genuine amber along as a comparison guide, when shopping for amber articles, is very useful. For fine details of inclusions, bubbles, etc., examine the pieces with a 5x or 10x magnifier.

Swirl lines in synthetic materials differ in appearance from swirls within true amber. This is the result of the different ways the substances were formed. Synthetic resins are poured into a mold, causing turbulent swirls. Amber resins slowly oozed from trees to accumulate in layers over a long period of time. Because of this, swirl lines in the imitation do not show the same flow or "movement" as in true amber. When a piece of genuine amber which possesses swirls within the mass is placed beside a synthetic piece containing swirls, the difference becomes plain.

Examination of the back of bracelet sections or other flat cabochon shapes is useful in detecting pieces that were formed in open molds. An irregular surface on the back signifies that synthetic resins solidified while exposed to the atmosphere. Careful examination may also reveal other mold marks that indicate the specimen was formed in a mold, rather than cut from a piece of rough. (See Table 9.1 for a comparison of the characteristics of amber and some of its common imitators.)

GLASS IMITATIONS

Glass imitations are usually easily detected because of the greater hardness of glass and its much greater weight. Glass beads are also cold to the touch, whereas amber feels warm, and differences in the lusters of glass and amber are just as easily distinguished. Glass beads are often faceted in a manner similar to that of the antique faceted amber, and the beads range in hue

Table 9-1. Characteristics of amber and common imitations.

Material	Specific Gravity	Refractive Index	Under Knife	Hot Point Odor
Amber	1.08	1.54	Splinters readily.	Resinous.
Copal	1.06	1.53	Splinters readily; softer than amber.	Resinous.
Phenol bakelites (bakelite, catalin, etc.)	1.26 to 1.28	1.64 to 1.66	Sectile, tough.	Acrid.
Urea bakelites (beetle, etc.)	1.48 to 1.55	1.55 to 1.62	Sectile, tough.	Acrid.
Caseins (galalith, lactoid, etc.)	1.32 to 1.43	1.49 to 1.51	Sectile, rather tough.	Burned milk.
Acrylate resins (perspex, diakon)	1.18 to 1.19	1.49 to 1.50	Sectile.	Burned fruit.
Polystyrenes (Trolutol, distrene)	1.05	1.59 to 1.67	Sectile.	
Slocum imitation	1.17	1.50 to 1.55	Sectile.	Sweet acrid, burned fruit.
Cellulose nitrate (celluloid)	1.37 to 1.43	1.49 to 1.52	Readily sectile.	Camphorous.
Cellulose acetate (safety celluloid)	1.29 to 1.35	1.49 to 1.51	Readily sectile.	
Bernat, Bernit	1.23	1.54	Readily sectile.	

from yellow to orange to brown. Yellow glass colored by uranium oxide will show a brilliant yellow-green fluorescence under ultraviolet light. Manganese is also used to produce yellow-colored glass, and this gives a dull green fluorescence quite different from the fluorescence of amber.[19]

Amateur collectors may test glass beads by simply rubbing them together. The hardness can be felt as the beads scratch against the hard surfaces of one another.

Another method used by "seasoned" collectors to detect an impostor of genuine amber is to gently bite the beads. Collectors swear that their teeth can "feel" the difference between the hardness of amber, glass and even ambroid or plastic. If the beads are amber, the material will be soft and brittle. Since ambroid is only slightly harder than amber, it may be difficult to detect using this unsophisticated method. Some say plastic and glass beads also have a different taste on the tongue. In any case, beginners using this bite test may want to practice on their own beads to get the "feel" of it.

HORN IMITATIONS

Another organic substance used to imitate amber is—surprisingly—horn. Horn beads are usually grayish tan in color, but may be dyed various colors. Small, opaque to translucent, yellowish barrel-shaped horn beads are im-

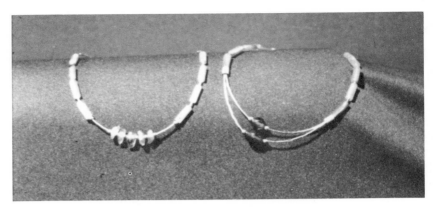

Fig. 9–10. Tube-shaped beads are made of horn.

ported from Ireland and occasionally passed off as amber. Figure 9–10 illustrates such beads fashioned into a necklace with amber and synthetic amber beads, but this is not a typical design in the imported material. More often, the beads are found strung as a rosary.[20]

Horn usually has a higher specific gravity than amber and sinks in a saturated solution of salt water (though an occasional bead may float). The hot point provides a useful test because horn emits an unpleasant odor, like that of burning hair, when touched with the red hot point.

The carved, claw-shaped watch fob in Fig. 9–11 resembles pressed amber and was identified as such in several examinations. Had it been tested with the hot point, the obvious odor of hair would have been detected. (See Fig. 9–12 for an illustration of a hot point tester.)

Fig. 9–11. Antique claw watchfob carved in China from horn, sold in an estate sale as amber. (A. Calomeni's collection.)

Fig. 9-12. Gem hot point testing instrument.

REFERENCES

1. Desautels, Paul E., *The Gem Kingdom*. New York: Random House, 1977, p. 64.
2. "Plastics," *Encyclopedia Americana, International Edition* **XXII:** 221-228. New York, 1975.
3. "Plastics," *Encyclopaedia Britannica* **XVIII:** 7-8. Chicago, 1971.
4. Allen, Jamey, "Amber and Its Substitutes, Part 3: Is It Real? Testing Amber," *The Bead Journal* **3,** *1:* 22, 1976.
5. Webster, Robert, *Practical Gemology*. London: N.A.G. Press, 1976, p. 182.
6. Anderson, B. W., *Gem Testing*. New York: Emerson Books, 1973, pp. 210-211.
7. Webster, Robert, *Gems, Their Sources, Descriptions and Identification*. Hamden, Connecticut: Archon Books, 1975, p. 514.
8. "Plastics," *Encyclopedia Americana* **XXII:** 222
9. "Plastics," *Encyclopaedia Britannica* **XVIII:** 7-8.
10. Webster, *Gems,* p. 514.
11. Liddicoat, R. T., Jr., *Handbook of Gem Identification*. Los Angeles: Gemological Institute of America, 1969, pp. 214, 223.
12. Anderson, *Gem Testing,* p. 212.
13. Webster, *Practical Gemology,* p. 141.
14. Anderson, *Gem Testing,* p. 212.
15. Webster, *Gems,* p. 514.
16. Webster, *Practical Gemology,* p. 141.

17. Allen, "Amber and Its Substitutes, Part 3: Is It Real? Testing Amber," *The Bead Journal* **3,** *1:* 20–31.
18. Webster, *Gems,* p. 515.
19. *Ibid.,* p. 516.
20. Allen, "Amber and Its Substitutes, Part 3: Is It Real? Testing Amber," *The Bead Journal* **3,** *1:* 22.

10
The Preparation and Working Of Amber

CLARIFICATION

That cloudy amber could be clarified apparently was known as early as the first century, for Pliny, in his *Natural History,*[1] mentioned that pieces of amber were "dressed by being boiled in the fat of a suckling pig by Archelaus, King of Cappadocia." This is thought by Willy Ley to be the first attempt at clarifying amber. Andreas Aurifaber made observations about clarifying amber in 1572, as did Johannes Wigand in 1590. It was the latter who discovered other oils could be used rather than the "fat of suckling pigs,"[2] and, by the end of the seventeenth century, methods for ridding amber of internal obscurations had been well established.

According to Bauer,[3] the method followed at the turn of the last century was as follows:

The rough material is completely immersed in rapeseed oil in an iron vessel, and then very slowly heated to about the temperature at which the oil boils and begins to decompose. It is then allowed to cool and this must take place just as slowly and gradually as the preliminary heating otherwise the clarified amber will become cracked and possibly fractured. The smaller the fragments of amber operated upon the quicker is the process completed; the heating of large pieces must be continued for the considerable period, and not infrequently needs to be several times repeated. The time required for the operation depends also upon the character of the material, for different pieces of amber of the same size will not require the same length of time to complete the operation. The clarifying process begins on the surface and spreads gradually inwards.

To understand how this process works, it must be remembered that minute air bubbles give amber its turbid appearance. When these air spaces are filled with oil of the same (or nearly the same) refractive index as amber, the material appears transparent, since the light is then allowed to pass

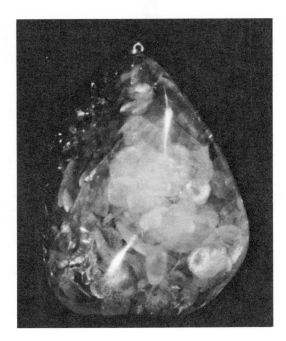

Fig. 10-1. Elegant ''sun-spangle'' fissures add to the radiating beauty of the shining golden amber. *(Courtesy Amber Forever, Florida.)*

Fig. 10-2. ''Sun-spangle'' or explosion inclusions can be light or can have a brown ''toasted'' appearance.

through the material without interference. The refractive index of rapeseed oil, a light yellow oil obtained from pressing rapeseeds, is 1.475, which is close to that of amber, so this is the oil most often used. Coloring can be mixed with the oil as desired. The discoidal fissures, or "sun-spangles," in amber are believed to be caused by droplets of trapped water which flatten to circular forms. When amber is clarified, great care is taken to make the fissures conspicuous rather than to produce a completely transparent piece. If properly done, the fissures assume an elegant appearance and display characteristic radiating circular forms within the transparent mass. These fissures are considered very attractive by amber connoisseurs. The presence of "sun-spangles" with brown edges distinguishes clarified from natural transparent amber. (See Color Plates 43–44.)

HEAT TREATMENT

To produce a rich brown "antique" color, amber is imbedded in pure sand in an iron pot, then increasingly heated for 30 to 40 hours.[4] The resulting rich brown color closely resembles the oxidized amber hues produced by aging. Such treated amber is often sold under the term "antiqued amber." (See Color Plates 48 and 50 for examples of "antique" amber.)

MODERN METHODS OF TREATMENT

According to the catalog of the Soviet firm, Almazjuvelirexport, "one kind of yellow cloudy amber is not as desirable as other varieties. Therefore, a hardening process has been developed. The color is removed in an autoclave, then the amber is hardened in electric ovens. The amber becomes beautiful in color and markings."[5] This meager description is the most up-to-date explanation obtainable from current amber wholesale firms relating to improving color and clarity of amber. The jewelry trade, for obvious reasons, does not provide details of the processes used, but these must be along the lines of the treatments described below.

An unheated autoclave or "pressure cooker" is used to drive the oil into air pockets of the porous material to produce clarification. Next, the amber is placed in an electric oven and gradually heated to soften it. During this stage, any air bubbles left within the mass of amber expand to form discoidal fractures. At the same time, heating speeds the process of oxidation and causes the amber to deepen in color. Oxidation occurs from the outside in, causing outer layers to be generally darker than the cores. The longer amber is heated, the darker it becomes. After porous pieces have been subjected to this treatment, they tend to be harder and more durable than they were previously. The process produces a beautiful amber, in great demand by amber connoisseurs, and one which is still considered to be genuine amber. It is most commonly found in the finished pieces imported into the United States from manufacturers in Poland, Russia and West Germany.

Fig. 10–3. Sun-spangle amber necklace made in Poland. (Z. Ladak's collection.)

AMBROID: PRESSED OR RECONSTRUCTED AMBER

Around 1880, in Vienna, it was discovered that small pieces of amber could be fused together, by the use of heat and pressure, into large blocks of so-called "pressed amber" or "ambroid." By 1881, techniques used in the German amber industry had been refined to make the process commercially feasible.

Irregular and small pieces of amber, otherwise unusable in the manufacture of jewelry or art ornaments, as well as fragments left over from working amber articles, are the raw materials for pressed amber.

Pieces are carefully selected according to size, and the weathered crusts are scraped and cleaned to remove all impurities. When exposed to a temperature of 170 to 190 degrees centigrade, the amber softens to a rubber-like consistency without disintegrating. To prevent decomposition during the process, the amber is placed in a deep steel tray with a perforated partition, heated to about 200 to 250 degrees centigrade and air is excluded. Hydraulic pressure of about 50,000 pounds per square inch is applied to force the soft amber through the perforations into a second compartment, where it cools into a solid flat block. A variation of this method forces the softened amber through perforated plates into molds of desired shapes and sizes.

Passing the softened amber through perforated plates causes fragments to lose their shape and intermix, with the result that the finished product tends to be less clear and brilliant than non-treated amber and has a hazy or "cirrus cloudy" appearance. Ambroid is also harder than natural amber.

A method known as "Spiller" was used in the past to form pressed amber, and is still occasionally used today. Amber chips were evenly heated in a steel mold with an airtight, sealed cover; then hydraulic pressure was applied to the cover. With no means of escape, the amber welded together into a solid mass. However, when such material is examined under magnification, the hazy outlines of the fused chips can still be seen. Pressed amber formed in this manner is generally a transparent dark brown color because of the darkening caused by heat, but some pieces retain their original cloudy yellow appearance and produce an uneven coloration in the block.

In a 1913 issue of the *Mining Journal,* E. Bellmann[6] defined compressed amber or ambroid as "amber produced from small pieces without admixture of foreign substances, fused by heat under high pressure in a vacuum. According to the size and color of the material used, the strength and duration of the pressure and the height of the temperature, the color shades including the characteristic cloudy spots of raw amber can be reproduced in the compressed article. Other characteristic properties of the natural 'stone' suffer no alteration providing the crude material used does not contain any impurities whatever, such as dirt, dust, weathered cortex, or foreign substances inside the lumps. To free the lumps thoroughly from the cortex is very tedious and slow work, in which about 150 girls in the factory and 500 female home-workers are constantly engaged, partly in Palmnicken and partly in Königsberg. The amber is pressed into plates, bars, cones, and of various shapes and measurements."

Early pressed amber products had one major drawback: they did not retain their original color and transparency, but became cloudy with age. "The cirrus clouds, which appeal to some in amber and appear in the process of manufacturing amberoid, became, after a few months, unpleasant in appearance, while articles made from the transparent amberoid lack the luster and warmth of these made from a natural mass because the structure reveals undulating lines."[7] The flow lines referred to give the appearance of water and glycerine flowing together. Such may be seen under magnification in transmitted light, and this roiled appearance indicates pressed amber. When a pressed piece is placed beside natural amber, these appearances become readily apparent. Bubbles in ambroid are more elongated and somewhat flattened, as compared to those in natural amber, which retain round or spherical shapes.[8] Also in contrast to natural amber, pressed material tends to soften in ether, but the refractive index and specific gravity are similar to those of natural amber. Polariscopic examination reveals an even light appearance when rotated between darkened polaroids.[9] (See Table 10–1.)

Because amber naturally darkens, becoming a pleasing reddish color, after being polished and exposed to the atmosphere, pigments are sometimes

Table 10-1. Ambroid characteristics.

Refractive index	1.54
Specific gravity	1.04 to 1.08
Solubility	Softens under ether.
Polariscope, dark position	Even light appearance in all positions.
Microscopic examination	a. Roiled appearance, undulating lines.
	b. Elongated air bubbles.
	c. Texture varies from uniform mistiness to hazy outlines of fused pieces.

added to pressed amber to provide deeper shades suggestive of age. Green hues have also been imparted. Color Plate 49 illustrates a bracelet of green pressed amber, as contrasted to the natural yellow amber slab on which it is resting.

Except for addition of minute quantities of color, pressed amber contains no foreign substances. When the process was first used, ambroid was more expensive than natural amber, but this is no longer the case. From the time the Royal Amber Works in Königsberg purchased the ambroid industry, around the beginning of the twentieth century, pressed amber was considered especially valuable for smoker's articles because it was more hygienic than wood, horn, bone or celluloid, and was also reputed to possess curative powers of its own. Ambroid was excellent, too, for making large art objects. Königsberg remained the center for manufacturing pressed amber products until World War II. After the amber area of Samland became incorporated into the Soviet Union, ambroid began to be made in Poland as well as in Germany.

The Soviet catalog of amber exports designates the yellow, uniformly colored cloudy material as "pressed" amber (see Color Plate 46), while transparent brown with swirls of darker material throughout the mass is designated as "reconstructed" amber. More often, the terms "pressed" and "reconstructed" are used interchangeably and are applied to any product formed by the process explained above.

Recently, techniques in processing ambroid have been developed so that sun-spangle inclusions can also be expertly induced into pressed amber. With close examination of the fissures, using a jeweler's eyeloop, they appear unnatural, without the distinct circular or discoidal outlines that are so characteristic in natural amber.

In summary, the hallmarks of pressed amber are a roiled appearance throughout or along junctions between the separate amber pieces, sometimes only visible under magnification, and a tendency to soften under ether. Insect inclusions have been introduced into pressed amber, but may be distinguished by their flattened appearance, rather than the fullness

Fig. 10-4. Pressed amber pendant with sun-spangle inclusions. Note the indistinct shape of the spangles.

which characterizes natural insect inclusions. Such inserted insects are whole, for the most part, and lack the dismemberment which is so often seen in natural amber, where insects struggled to free themselves, losing legs or wings in the process.

EARLY METHODS OF WORKING AMBER

Amber artifacts from Stone Age graves in the Baltic region show that the primitive inhabitants were familiar with methods for crudely shaping and finishing many kinds of small objects. Prior to the advent of metals, such shaping was accomplished by the use of sandstones or other gritty stones which served as saws and files. Wood, supplied with fine sand, could be used for smoothing surfaces. Holes probably were drilled, as in other early cultures, by the simple device of twirling a pointed stick pressed against the amber, the point of which was supplied with a slurry of water and fine sand as an abrasive. Wood was also used for polishing, and when supplied with wood ashes or some other very fine abrasive, could readily impart a polish to amber.

With the appearance of bronze (and later, iron) in the Baltic region, more sophisticated work was achieved, but the methods for shaping, smoothing and polishing remained little changed over the centuries. As late as the seventeenth century, natives of the region worked amber, especially during long winter evenings when there was little else to do, with the most primitive, yet effective, implements. The stages through which the amber was processed consisted of removal of the outer, soft, cracked and discolored crust, generally by scraping; then sawing larger pieces into smaller ones as required; shaping the pieces by the use of scrapers and files; and, lastly,

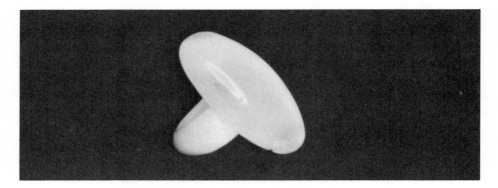

Fig. 10–5. Amber ring carved by peasants in Poland.

smoothing and polishing. Where metal files were unavailable, gritty stones of various types could be used, lubricated with water to take away the amber particles and dust. Polishes were obtained by smearing a paste of wood ashes or chalk on pieces of wood and rubbing the wood briskly over the surfaces of the amber. By such simple means, peasants were able to produce large round or oval beads, bars for brooches, finger rings, medallions and even crucifixes. A popular carved object was the heart, which, in addition to its obvious ornamental value, was believed to confer protection upon its wearer. Amber buttons were made in a variety of shapes—round, square, triangular, etc. Some even were shaped into representations of flowers. Finger rings, shank and all, were carved from single pieces, but at other times, such rings were made in two pieces: the ring itself, and a "stone" attached to the top. Products of such cottage industries reached their highest development in Poland during the last century in the districts of Kaszuby

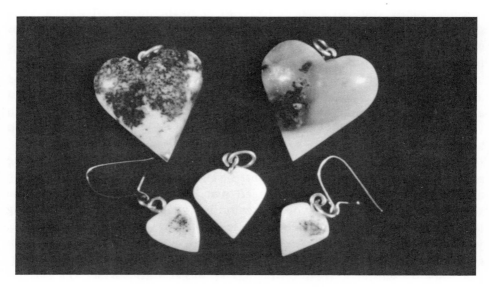

Fig. 10–6. Hand-carved amber hearts from Poland.

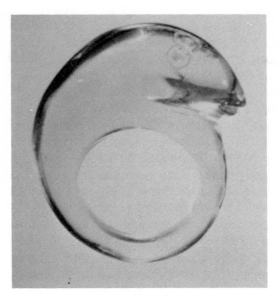

Fig. 10–7. Hand-canrved amber ring from the Dominican Republic.

and Kurpie, in the Bory Tucholskie Forest in Pomerania, and in the Paszcza Myzsyniecka Forest in the Narew River basin. Amber pits were worked in these regions until the middle of the last century, providing a local source of raw material.

Though entirely hand-made objects were produced from amber for centuries, it was not until the eighteenth century that a kind of mechanization was introduced; in this case, the adaptation of a spinning wheel into a crude lathe for turning round and cylindrical objects from amber. The amber lump was attached to the axle and the latter turned with a foot treadle, while the shaping was accomplished by placing a piece of broken glass against the amber. As may be expected, this crude lathe was soon superseded by more sophisticated devices, and the so-called "spinning wheel lathe" is now seen in use only in a few isolated areas in the Kurpie district of Poland.[10]

Amber carvings are still produced on a small scale in Germany, Poland and in the Lithuanian SSR, as are pictures made in mosaic fashion. However, such objects are few in number and generally are available only in the Lithuanian SSR or in the Soviet Union, very few being exported. In regard to mosaics and the related inlay, thin pieces of amber are cut to fit together like the pieces of a jigsaw puzzle, and by clever choice of colors and textures, geometrical designs or even pictorial designs or scenes are created. After all pieces are fitted together and cemented into place, the whole slab is smoothed and polished at one time. Such finished pieces in the past have been used for panels for walls or cabinets and to cover the side of a box or jewelry casket. Other materials of about the same hardness as amber were also used in combination with amber to provide greater versatility in design. For example, ivory and tortoise-shell were used, especially in the manufacture of attractive chessboards.

INDUSTRIAL PRODUCTION OF AMBER OBJECTS

In the early 1900's, the prevalent belief that amber possessed germicidal properties led to a vast amount being used in the manufacture of smoker's articles. Nearly one-half of the total production in the Baltic region was devoted to the making of cigar and cigarette holders and pipe mouthpieces. Vienna alone used up to 40 percent of the annual yield of East Prussian amber, and its factories were world-famous for the high quality of such goods produced in that city. Smoker's articles were also made in Königsberg (Kaliningrad) and Memel (Klaipeda). Many mouthpieces and holders were made on lathes and pierced with machine drills, but some special shapes were made by sawing and hand-shaping. Pipe stems were made from slender amber strips which were drilled and then immersed in hot linseed oil to render them clearer and pliable enough to be bent into the curved forms desired for mouthpieces.

Other articles made from amber, such as small carvings and beads, were specialties of factories in Danzig (Gdańsk), Berlin and Stolp (Slupsk), and by the turn of the last century, the latter had become world suppliers, tailoring their products to suit the special requirements of each country. Catalogs of manufacturers list six styles of beads from which to choose:

Standard Styles of Beads[11]

1. Olives	Elongated elliptical beads.
2. Zotten	Barrel-shaped, cylindrical; slightly rounded, almost plane at the two ends.
3. Grecken	Like zotten, but shorter.
4. Beads proper	Spherical.
5. Coral	Faceted beads.
6. Horse-corals	Flat, clear beads faceted at the two ends.

During the early 1900's, transparent golden "corals" (or faceted beads) were in great demand in the United States. These beautiful hand-crafted beads can be found in antique shops today in a golden amber color. In Europe, however, opaque bastard amber in the shape of olive or round beads was generally considered the most valuable, while France preferred water-clear beads. China and Korea produced round beads for mandarins and, in 1904, China imported approximately $25,000 worth of rough amber for this purpose alone.

CUTTING AND POLISHING AMBER

Most important in working amber is selecting rough material suitable for producing the desired finished product. Each piece of raw amber should be carefully considered in the design so natural inclusions and even portions of

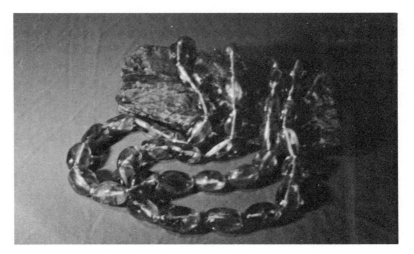

Fig. 10–8. Hand-turned olive-shaped beads from Poland. (H. Evens' collection.)

Fig. 10–9. Hand-carved amber bug made in Lithuania (actual size: 2½ inches long).

the crust may be incorporated into the detail of the object produced. The crust may be ground off to display the true amber colors beneath the surface, or portions of the crust may be left untouched to provide a deeper coloration to such surfaces.

Shelly amber, formed in·concentric layers, is weakly adherent along layer boundaries and, if drilling is required, the possibility of fracturing is more likely when the piece is drilled in the same plane as the layers. Rough pieces of this variety are better used in larger blocks requiring little drilling.

HAND METHODS

Many artists work amber using simple hand tools for cutting, shaping and polishing. The first step is to decide on the shape of the finished piece and to cut any large unneeded material away with a jeweler's saw or hacksaw, using slow, even sawing strokes. Great care must be taken to avoid any wobbling or bending of the saw, as pressure on the amber piece may cause it to chip or perhaps fracture completely.

After the desired shape has been roughly cut out, files are used to further shape the piece.

A coarse rasp file may be used to remove thick crust, but its use is only recommended when thick material needs to be quickly removed. Generally, it is safer to use flat files in several degrees of coarseness for removing material and smoothing the surface. Begin shaping by making even strokes with the file and crossing these by strokes at right angles to eliminate scratch lines. Files readily clog as work proceeds, and a file card (file cleaner) should be kept handy and used frequently to keep the teeth clean. Small needle files of different shapes are employed to remove material from depressions or other irregularities in the surface, and a small triangular file is particularly useful. However, care must be taken to avoid the wedging of small files in crevices, as this easily causes chipping or even results in the whole piece splitting in two. After the shape is roughed out with files, the piece is ready for sanding.

All surfaces must now be further smoothed to remove all file scratches. For this purpose, sandpapers of decreasing coarseness are used and prepared as follows. Make the first of several boards with coarse grit papers, such as sandpaper 220, emery paper 220 or coarse grit aluminum oxide paper. Tack the sandpaper around an 8 by 10 by 1/2 inch board. Holding the piece of amber in the fingers, rub the lump over the sanding board, using a scrubbing circular motion and varying movements so as to sand all surfaces of the piece. Repeat this process, using a medium (400) grit sanding board. The final sanding step requires fine grit aluminum oxide paper or sandpaper 600.

When working any of the ambers or fossil resins, hand-sanding can be done dry, since the work is generally slow and heat does not soften the amber. However, copal, or the so-called "African amber," is far more brittle and softer than the fossil ambers and, for this reason, should be sanded wet, using a waterproof sandpaper. Also, the lower melting point of copal results in the material becoming gummy during dry sanding; wet sanding prevents both temperature rise and excessive clogging of the sandpaper.

Upon completion of sanding, closely examine the piece to be sure that all fine scratches are removed before polishing. It is very exasperating to polish a specimen and then discover one or two deeper scratches that were overlooked. If this happens, go back to sanding until such scratches are eliminated and the entire surface has a uniform frosted appearance.

The last step is polishing, and various suggestions by different professional lapidaries have been made regarding methods for buffing and polishing amber. Some use tripoli or rottenstone as the polishing agent, with oil as the lubricant; others recommend aluminum oxide, tin oxide, Linde A or rouge. Tin oxide and Linde A may be applied as a paste made up with water, whereas rouge is applied dry. In Mexico, it is reported that dry cigarette ashes are used as a polishing agent. Whichever agent is chosen, the following instructions serve as a general guide for the polishing procedure.

Begin by dipping the piece into the lubricant, then into the polishing compound. Rub vigorously on a smooth surface, such as a leather razor strop or a polishing board made by tacking a piece of chamois to a 3 by 7 inch board. Holding the amber in the fingers, rub all surfaces over the polish board to develop the finish. Clean off the piece from time to time and examine for places which are unevenly polished. Hollow or depressed areas require wrapping dowels with chamois to reach the surfaces to be polished.

MECHANICAL POLISHING

Amber can be polished using power-driven lapidary equipment, but because of the fragile nature of amber as compared to mineral gemstones, much more care is required during all stages. Amber will not stand any kind of high-speed friction grinding, and some specimens under a state of strain may shatter. Overheating from too much friction causes upper layers to melt and produces a rippled or "orange peel" surface. If the amber approaches its melting temperature for an excessive period, volatile substances escape and the piece is left weakened and more brittle. For these reasons, when power-driven wheels or laps are used, the speed should be slowed to around 100 RPM, and large amounts of water should be supplied to abrasive or polishing wheels to avoid heating, Such reduced speed is especially important when shaping pendants and cabochons. Working on several pieces at one time may be necessary because the pieces build up static electricity charges, which can be felt as tremors in the hand. If this continues, the pieces may shatter. By alternating from one piece to another, the electrical charge dissipates. Felt wheels are excellent for the finish polish, with any of the polish agents mentioned previously, but with water as the lubricant. Never use a felt wheel that has been used for polishing hard minerals, since scratches may develop on the amber.[12] Avoid using diamond laps, as they cause the surface to become dull and cloudy, giving an unpleasant appearance. Linde A, water and a felt wheel rotating at a slow speed provide a combination that produces a finished piece with a fine luster.

DRILLING AMBER

Amber may be drilled with a hand drill or an electric bead drill. When a hand drill is used, hold the amber object securely in te fingers to avoid splitting the piece. Using a vise on amber causes chipping and is not recommended. A hand drill may be used dry, as the speed can be varied to avoid excessive frictional heating. If an electric drill is used, however, a drill stand or commercially designed bead drill rig is necessary, because the amber must be submerged under water to avoid overheating while being drilled. Commercial bead drills not only have a stand, but also are designed so the drill bit can be moved in and out of the piece frequently, thus avoiding overheating. It is important, of course, not to attempt to hand-hold an electric drill while perforating a bead that is submerged in water!

When piercing, it is good practice to drill only halfway through the

Fig. 10–10. Commercial bead drill for perforating stone materials submerged in water.

Fig. 10–11. Amber sculpture made in Lithuania (actual size amber lump: 2½ inches tall).

material from one side, then turn the piece over to complete the hole. This prevents chipping the amber as the drill breaks through. If the drill clogs or sticks in the amber, add oil to the bit to loosen it. Tugging or twisting usually results in chipping.

FACETING AMBER

Amber is easily faceted if the lap speed is reduced. Vargas[13] suggests the use of a wax lap with copious amounts of lubricant, and Linde A as the polish. Cutting may be slow, resulting from clogging of the lap with amber particles. The use of more water helps flush away such debris. Vargas provides the following information for faceting amber gems in the manner used for mineral gemstones; that is, in brilliant or step-cut styles.

Cutting Angles	Cutting Lap	Cutting Speed
Culet — 43°	Fine — extra fine	100 RPM
Crown — 42°		

Polishing
Lap — wax
Speed — normal
Agent — Linde A

MAKING YOUR OWN "POLYBERN"

The embedding of small chips, nodules and grains of raw amber into plastic or synthetic resin is a process easily accomplished by the amateur to create attractive beads or other ornaments. This method makes possible the utilization of waste material or pieces too small for other purposes, and the material can be readily molded into attractive ornaments. Synthetic resin compounds can be purchased at craft stores and hobby shops and are cast in molds of polyethylene, glass or ceramic. For any casting plastic, be sure to follow the manufacturer's directions, as each type of resin requires a slightly different procedure. Practice with some inferior amber before attempting work with more expensive pieces.

The following procedure is similar to that used in the commercial production of "polybern" in the Baltic States of the Soviet Union, Poland and other areas where this type of processed amber product is made.

1. Select amber chips you wish to embed according to the size of your mold. Small chips are best. Make sure they are free of dirt and sand.

2. The mold should be clean and have a highly polished wax finish.

3. Mix the resin in disposable paper cups, adding a yellow pigment, if desired, to match the color of the amber. Do not inhale fumes or let the plastic touch the skin.

4. Pour a thin layer of plastic resin into the mold and allow it to set.

5. Add a layer of fresh plastic resin to the top of the first layer, then embed the amber chips. Allow this layer to harden before proceeding.

6. Fill the mold with a third layer of resin and leave undisturbed until the plastic solidifies. Follow the manufacturer's instructions for removal of the casting from the mold.

REFERENCES

1. Pliny, *Natural History, Book 37,* translated by D. E. Eichholz. London: William Heinemann, 1962, p. 199.
2. Ley, Willy, *Dragons in Amber.* New York: The Viking Press, 1951, p. 38.
3. Bauer, Max, *Precious Stones* II: 538. London: Charles Griffin and Co. 1904 (reprinted by Dover Publications, 1968).
4. Williamson, George C., *The Book of Amber.* London: Ernest Benn 1932, p. 233.
5. Almazjuvelirexport, *Bernstein.* Moscow, 1978, p. 1.
6. Bellmann, E., "Recovery and Treatment of Amber at Palmnicken (East Prussia)," *The Mining Journal* 102: 722, July 1913.
7. Anonymous, "Amber," *Manufacturing Jewelers* 92, *3:* 21, February 1933.
8. Anderson, B. W., *Gem Testing.* New York: Emerson Books, 1973, p. 323.
9. Liddicoat, R. T., Jr., *Handbook of Gem Identification.* Los Angeles: Gemological Institute of America, 1969, p. 327.
10. Zalewska, Zofia, *Amber in Poland, A Guide to the Exhibition.* Warsaw: Wydawnictwa Geologiczne, 1974, p. 98.
11. Bauer, *Precious Stones* II: 547.
12. Anonymous, "Practical Shop Notes," *The Jeweler's Circular* January 9, 1915.
13. Vargas, Glenn and Martha, *Faceting For Amateurs.* Palm Desert, California, 1969, p. 259.

ADDITIONAL REFERENCES

Alexander, A. E., "Renewed Interest in Amber, Organic Compound, Cited, the Gemologists' Corner," *The National Jeweler,* March 1976, p. 27.

Borglund, Erlund and Flauensgaard, Jacob, *Working in Plastic, Bone, Amber and Horn.* New York: Reinhold, 1968.

Hunger, Rosa, *Magic of Amber.* London: N.A.G. Press, 1977.

Kennedy, Gordon *et al., The Fundamentals of Gemstone Carving.* San Diego, California: Lapidary Journal, 1967, pp. 68–73.

Rath, Muriel, "Golden Amber, The Magnificent Historian," *Lapidary Journal* **25,** *1:* 36–46, April 1971.

Sinkankas, John, *Van Nostrand's Standard Catalog of Gems.* New York: Van Nostrand Reinhold, 1968, p. 79.

Bibliography

Alexander, A. E., "Renewed Interest in Amber, Organic Compound, Cited, 'The Gemologist's Corner,' " *The National Jeweler,* March 1976, p. 27.

Allen, Jamey D., "Amber and Its Substitutes, Part 1: Historical Aspect," *The Bead Journal,* Winter 1976, pp. 15–19.

Allen, Jamey D., "Amber and Its Substitutes, Part II: Mineral Analysis," *The Bead Journal* **2,** *4:* 11–12, Spring 1976.

Allen, Jamey D., "Amber and Its Substitutes, Part III: Is It Real? Testing Amber," *The Bead Journal* **3,** *1:* 20–31, Winter 1976.

Almazjuvelirexport, *Bernstein.* Moscow, 1978.

Anderson, B., *Gemstones for Everyman.* New York: Van Nostrand Reinhold, 1971.

Anderson, B. W., *Gem Testing.* New York: Emerson Books, 1973.

Andrée, Karl, Öst Preussens Bernstein und Seine Bedeutung. *Östdeutscher Naturwart* **S3:** 183–189, 1924.

Andrée, Karl, "Öst Preussens Bernstein und seine Bedeutung," *Östdeutscher Naturwart* **S3:** 120–134, 1925.

Andrée, Karl, *Der Bernstein und seine Bedeutung in Natur und Geisteswissenchaften, Kunst und Kunstgewerbe.* Technik, Industrie und Handel. Königsberg, 1937.

Andrée, Karl, *Der Bernstein, das Bernsteinland und sein Leben.* Stuttgart: Kosmos Booklet, 1951.

Arem, Joel, *Gems and Jewelry.* New York: Bantam Books, 1975.

Arem, Joel, *Color Encyclopedia of Gemstones.* New York: Van Nostrand Reinhold, 1977.

Aycke, John, Chr. von, *Fragmente Zur Naturgeschichte Des Bernsteins.* Danzig, 1835.

Axon, G. V., *The Wonderful World of Gems.* New York: Criterion Books, 1967.

Baer, John W., "Floating 'Gold' from the Baltic," *The Jeweler's Circular-Keystone,* October 1937, pp. 82–86.

Bajeri, Johann Jacobi, *Oryktographia Norica. . .* Noribergae, 1708.

Barrera, Madame De, *Gems and Jewels, Their History, Geography, Chemistry and Analysis.* London: Richard Bentley, 1860.

Barry, T. Hadley, *Natural Varnish Resins.* London: Ernest Benn, 1932.

Bauer, Jaroslav, *A Field Guide in Color to Minerals, Rocks and Precious Stones.* London: Octopus Books, 1974.

Bauer, Max, *Precious Stones* **II.** London: Charles Griffin & Co., 1904 (reprinted by Dover Publications, 1968).

Beard, Alice and Rogers, Frances, *5000 Years of Gems and Jewelry.* New York: Frederick A. Stokes, 1940.

Beck, Curt W., "The Origin of the Amber Found at Gough's Cave, Cheddar, Somerset," *Proceedings, University of Bristol Spelaeological Society* **10,** *3:* 272–275, 1965.

Beckmann, Franz, *Bedeutung Des Bernsteinnamens Elektron.* Braunsberg, 1859.

Bellmann, E., "Amber," *The Mining Journal* **101:** 129, 1913.

Bellmann, E., "Recovery and Treatment of Amber at Palmnicken (East Prussia)," *The Mining Journal* **102:** 122, 1913.

Berendt, G., *Der im Bernstein befindlichen Organischen Reste Der Vorwelt.* Berlin, 1845.

Berendt, G., *Geology of Kursches Haff* [in German], 1869.

Berry, E. W., "The Baltic Amber Deposits," *The Scientific Monthly* **24:** 268–278, 1927.

Berry, E. W., "The Past Climates of the North Polar Region," *Smithsonian Miscellaneous Collection 82, No. 6,* 1930.

Block, J. Anderson, *The Story of Jewelry.* New York: William Morrow, 1974.

Bock, Friedrich S., *Versuch einer kurzen Naturgeschichte des Preussischen Bernstein. . .* Königsberg, 1767.

Bolsche, Wilhelm Von, *Im Bernstein Wald.* Berlin, 1927.

Borglund, Erlund and Flauensgaard, Jacob, *Working in Plastic, Bone, Amber, and Horn.* New York: Reinhold, 1968.

Bradford, Ernle, *Four Centuries of European Jewellery.* Middlesex, England: Spring Books, 1953.

Brill, Robert (Ed.), *Science and Archaeology.* Cambridge: M.I.T. Press, 1971.

Brouwer, P. A., *Economica Dominicano* **11,** *16,* 1959.

Brues, Charles, "Insects in Amber," *Scientific American* **185,** *5:* 56–61, 1951.

Buddhue, J. D., "Mexican Amber," *Rocks and Minerals* **10:** 170–171, 1935.

Budge, E. A. Wallis, *Amulets and Talismans.* New York: University Books, 1961.

Buffum, Arnold, *The Tears of the Heliades or Amber as a Gem.* London: Sampson Low, Marston and Co., 1897.

Butenas, Petras, "Gintaro Sneka," *Karys* **4:** 110–114, Balandis, 1973.

Butenas, Petras, "Gintaro Sneka," *Karys* **5:** 159–164, Geguže, 1973.

Cammann, S., *Substance and Symbol in Chinese Toggles.* Philadelphia: University of Pennsylvania Press, 1962.

Carpenter, F. M., "Fossil Insects in Canadian Amber," *University of Toronto Studies, Geological Series* **38:** 69, 1935.

Carpenter, F. M., Flosom, J. W., Essig, E. O., Kinsey, A. L., Brues, C. T., Boesel, M. W., and Ewing, H. E., "Insects and Arachnids from Canadian Amber," *University of Toranto Studies, Geological Series No. 40,* 1937.

"Amber-Substance of the Sun," *Chemistry* **45:** 21–22, 1972.

Chhibber, H. L., *The Mineral Resources of Burma.* London: Macmillan & Co., 1934.

Childe, Gordon Vere, *The Dawn of European Civilization, 3rd Ed.* New York: Knopf, 1939.

Cockerell, T.D.A., "Some American Fossil Insects," *Proceedings of the U.S. National Museum* **51:** 89, 1916.

Conwentz, H., *Die Flora des Bernsteins.* Bd. 2, Die Angiospermen des Bernsteins. Danzig, 1886.

Conwentz, H., "Der Bernsteinfichte," *Ber. Deutsch Bot. Ges.* **4:** 375–377, 1886.

Conwentz, H., *Monographie der baltischen Bernsteinbäume.* S. 151, Taf. 18, Danzig, 1890.

Cook, John, M.D., *The Natural History of Lac, Amber, and Myrrh.* London, 1770.

Cuba, Johannis de, *DeLapidibus. Hortus Sanitatis,* Cap. ixx. Strassburg: Jean Pryss, ca. 1483–1491.

Dana, Edward S., *Dana's Textbook of Mineralogy, 4th Ed.* New York: John Wiley & Sons, 1949.

d'Aulaire, Emily and Ola, "For Ever—Amber," *Reader's Digest,* Reprint from Scandinavian Edition, December 1974.

d'Aulaire, Emily and Ola, "Amber: Gold of the North," *Scanorama Magazine* January 1978, pp. 61–64.

Desautels, Paul E., *The Gem Kingdom.* New York: Random House, 1977.

Durham, J. Wyatt, "Amber Through the Ages," *Pacific Discovery* **10,** 2: 3–5, 1957.

Encyclopaedia Britannica. Chicago, 1971.

Encyclopaedia Britannica, Micropaedia (Ready Reference). Chicago, 1974.

Encyclopedia Americana, International Edition. New York, 1975.

Encyclopedia Lituanica. Boston, 1970.

Erikson, Joan Mowat, *The Universal Bead.* New York: W. W. Norton & Co., 1969.

Evans, Joan, *Magical Jewels of the Middle Ages and Renaissance.* New York: Dover Publications, 1976.

Farrington, Oliver C., "Amber," *Dept. Geology Leaflet No. 3.* Chicago: Field Museum of Natural History, 1923.

Fernie, W. T., *Precious Stones: For Curative Wear, and Other Remedial Uses.* Bristol: John Wright & Co., 1907.

Fielder, M., "What is this Gem Called Amber?" *Lapidary Journal* **30,** 5: 1244–1249, 1976.

Fielding, William J., *Strange Superstitions and Magical Practices.* Philadelphia: Blakiston, 1945.

Frondel, Judith W., "Amber Facts and Fancies," *Economic Botany* **22:** 4, October–December 1968.

Gemological Institute of America, *Manual for Colored Stones.* Los Angeles, 1971.

Gimbutas, Marija, *The Balts.* New York: Frederick A. Praeger, 1963.

Goeppert, H. R. and Berendt, G. C., *Die Bernstein und die in ihm befindlichen Pflanzenreste der Vorwelt.* Bd. 1, Berlin, 1845.

Goeppert, H. R., "Über die Bernsteinflora," *Monatsberichte König. Academic der Wissenschaften.* Berlin, 1853.

Goeppert, H. and Menge, A., *Die Flora des Bernsteins und ihre Beziehungen Zur Flora der Tertiärformation und der Gegenwart.* Bd. 1, Danzig, 1883.

Goldsmith, E., *Proceedings, Academy of Natural Science.* Philadelphia, 1879.

Green, Roger Lancelyn, *Myths of the Norsemen.* London: Bodley Head, 1962.

Gudynas, P. and Pinkus, S., *The Palanga Museum of Amber.* Vilnius: Mintis Books, 1967.

Haddow, J. G., *Amber, All About It.* Liverpool: Cope's Smoke Room Booklets, No. 7, 1891.

Harrington, B. J., "On the so-called Amber of Cedar Lake, N. Saskatchewan, Canada," *American Journal of Science* **42,** 3: 332–338, 1891.

Hartmann, P. J., *Succini Prussici physica et civilis historia.* Frankfurt, 1677.

Heer, Oswald, *Die Tertiäre Flora der Schweiz.* Bd. 3, Winterthur, 1859.

Heer, Oswald, *Miocän baltische Flora.* Königsberg, 1869.

Helm, Otto, "Mittheilungen über Bernstein. 2, Glessit, 3 Über sicilianischen und rumanischen Bernstein," *Schriften der Naturforschenden Gesellschaft zu Danzig.* N.F.5, Danzig, 1881, pp. 291–296.

Helm, Otto, *Schriften Naturforschenden Gesellschaft.* N.R., Nr. 2, Danzig, 1885, pp. 234–239.

Helm, Otto, "Mittheilungen über Bernstein. 14, Über Rumanit. . ., 15, Über den succinit und die ihm verwandten fossilen Harze," *Schrift. d. Naturf. Ges.* N.F.7, Danzig, 1891, pp. 189–295.

Helm, Otto, "On a new fossil, amber-like resin occurring in Burma, from Upper Burma," *Records of the Geological Survey of India* **25:** 180–181, 1892.

Helm, Otto, "Mittheilungen über Bernstein. 17, Über den Gedanit, Succinit und eine Abart des letztern, den sogenannten mürber Bernstein," *Schrift. d. Naturf. Ges.* N.F.9, Danzig, 1896, pp. 52–57.

Hodges, Doris M., *Healing Stones.* Iowa: Pyramid Publishers of Iowa, 1962.

Hornaday, W. D., "Kauri Gum Deposits of New Zealand," *Mining Press* **110:** 181–182, 1915.

Hunger, Rosa, *Magic of Amber.* London: N.A.G. Press, 1977.

Jakubowski, M., "Amber Fishers and Mines," *The Jeweler's Circular-Weekly,* December 23, 1914, p. 14.

Jewell, W. B., *Geology and Mineral Resources of Hardin County, Tennessee.* Tennessee Division of Geology, Bull. 37, 1931.

Jobes, Gertrude, *Dictionary of Mythology, Folklore and Symbols, Part 1*. New York: The Scarecrow Press, 1961.

John, J. F., *Naturgeschichte des Succins oder des sogenannten Bernsteins*. Köln: Theodor Franz Thiriart, 1816.

Kaunhoven, F. "Der Bernstein in Östpreussen," *Jahrbuch der Preussischen Geologisches Landesanstalt*. Bd. 34 T 2, Berline, 1913.

Kemp, Edwin C., *American Counsular Commerce Reports*. Danzig: American Consulate, April 27, 1925.

Kennedy, Gordon et al., *The Fundamentals of Gemstone Carving*. San Diego: Lapidary Journal, 1967.

Keferstein, Chr., *Mineralogia Polyglotta*. Halle, 1849.

King, Albert A., "Texas Gemstones," *Report of Investigations No. 42*. Texas State Geology Dept., 1961.

Klebs, R., *Der Bernstein*. Königsberg, 1880.

Klebs, R., *Der Bernsteinschmuck der Steinzeit von der Boggerei bei Schwarzort und anderen Lokalitäten Preussen*. Königsberg, 1882.

Komarow, W. L., *Proiskhozhdenie Rastenii*. Led. 7 Moscow, 1943.

Kraus, Edward H., *Gems and Gem Materials*. New York: McGraw-Hill, 1939.

Kunz, George Frederick, *The Curious Lore of Precious Stones*. New York: Dover Publications, 1971.

Langenheim, Jean H., "Amber, A Botanical Inquiry," *Science* **163**, 3: 1156–1169, 1969.

Langenheim, Jean H., "Present Status of Botanical Studies of Ambers," *Botanical Museum Leaflets*. **20:** 8. Cambridge: Harvard University Press, May 8, 1964.

Laufer, Berthold, "Historical Jottings on Amber in Asia," *Memoirs of the American Anthropological Association* **I**, 3. Lancaster, Pennsylvania, 1907.

Leach, Marchia, *Standard Dictionary of Folklore Mythology and Legend*. New York: Funk and Wagnalls, 1949.

Ley, Willy, "The Story of Amber," *Natural History,* May 1938, pp. 351–377.

Ley, Willy, *Dragons in Amber*. New York: The Viking Press, 1951.

Liddicoat, R. T., Jr., *Handbook of Gem Identification*. Los Angeles: Gemological Institute of America, 1969.

Lomonosov, Mikhail V., [*Complete Collection of Works* **V**, in Russian]. Moscow-Leningrad, 1954.

Lucas, A., *Ancient Egyptian Materials and Industries*. London: Edward Arnold and Co., 1934.

Lüschen, Hans, *Die Namen der Steine*. München: Ott Verlag, 1968.

MacFall, Russell P., *Gem Hunter's Guide, 5th Ed*. New York: Thomas Y. Crowell, 1975.

"Amber," *Manufacturing Jewelers* **92**, 3: 21, 1933.

Murgoci, George, *Gisements Du Succin De Roumanie avec un Aperçu Sur Les Résines-Fosiles*. Bucarest: L'Imprimèrie De L'Etat, 1903.

McAlpine, J. F. and Martin, J. E. H., "Canadian Amber—A Paleontological Treasure-Chest," *The Canadian Entomologist* **101**: 819–822, August 1969.

McDonald, Lucile Saunders, *Jewels and Gems*. New York: Thomas Y. Crowell, 1940.

Narune, J., *Ambarita (Amberella) Leyenda*. New York: Council of Lithuanian Women, 1961.

Noetling, F., "Preliminary Report on the Economic Resources of Amber and Jade Mines Area in Upper Burma," *Records of the Geological Survey of India* **25**: 130–135, 1892.

Pačlt, J. A., *A System of Caustolites. Tschermaks Mineralogische und Petrographische Mitteilungen* (drittle Folge). Bd. 3, H. 4, Wien., 1953.

Parsons, Charles J., *Practical Gem Knowledge for the Amateur*. San Diego: Lapidary Journal, 1969.

Pelka, Otto, *Bernstein*. Berlin: Richard Carl Schmidt, 1920.

Penrose, R.A.F., Jr., "Kauri Gum Mining in New Zealand," *The Journal of Geology* **20**: 1, January–February 1912.

Petar, Alice V., "Amber," *U.S. Bureau Mines Information Circular 6789*. 1934.

Pliny, the Elder, *The Natural History of Pliny* **10**, *Book 37,* Translation by D. E. Eichholz. Cambridge: Harvard University Press, 1962.

Pliny, the Elder, *The Natural History of Pliny* **6**, Chapter 11, Translation by Bostock, J. and Riley, H. T. London: Henry G. Bohn, 1857.

Prockat, Friedrich, "Amber Mining in Germany," *Engineering and Mining Journal* **129**: 305–307, 1932.

Raimundas, Sidrys, *Third Symposium on the Sciences and Arts.* Chicago: Draugas Lithuanian Newspaper, April 8, 1978.

Rath, Muriel, "Golden Amber, the Magnificent Historian," *Lapidary Journal* **25**, *1:* 36–46, April 1971.

Rebaux, M., [Natural Amber], *Annales de Chimie et de Physique*. Paris, 1880.

Reed, A. H. and Collins, Tudor W., *New Zealand's Forest King—The Kauri*. Wellington, Auckland: Consolidated Press Holdings, 1967.

Research Reporter, "Amber Substance of the Sun," *Chemistry* **45**, *21:* 21–22, April 1972.

Rice, Patty, "Amber is Back," *Gems and Minerals* **488**: 14–17, June 1978.

Rudler, F. M., "A Piece of Amber," *Science for All*. London: Cassell and Co., ca. 1890, pp. 213–218.

Runge, W., *Der Bernstein in Östpreussen*. Berlin, 1868.

Ruzic, Raiko H., "Amber in Chiapas, Mexico," *Lapidary Journal* **27**: 1304–1305, 1400–1406, 1973.

Samples, C. C., "Amber," *England Mining Journal* **80**: *250,* 1905.

Sanderson, Milton R. and Farr, Thomas H., "Amber with Insect and Plant Inclusions from the Dominican Republic," *Science* **131**: 1313, 1960.

Saunders, W. B., Mapes, R. H., Carpenter, F. M., and Elsik, W. C., "Fossiliferous Amber from the Eocene (Claiborne) of the Gulf Coastal Plain," *Geological Society of America Bulletin* **85**: 979–984, June 1974.

Savkevich, S. S., "State of Investigation and Prospects for Amber in U.S.S.R.," *International Geology Review* **17**; *8:* 919–923, August 1975.

Savkevich, S. S. and Popkova, T. N., "New Data on 'Amber' From the Right Bank Area of the Kheta and the Khatanga Rivers," *Mineralogy, Doklady Akademii Nauk S.S.S.R.* **208**; *1–6:* 131–132, 1973.

Schweisheimer, W., M.D., "The Medical Power of Jewels,"*Jeweler's Circular-Keystone* **139**, *7:* 70–71, 93, 96, 1968.

"Petrified Tree Tears," *Science Digest,* August 1973, pp. 86–87.

Sendelio, Nathanaele, *Historia Succinorum. . .* Elbing, 1742.

Sinkankas, John, *Gemstones of North America*. New York: Van Nostrand Reinhold, 1966.

Sinkankas, John, *Gemstone and Mineral Data Book*. New York: Collier Books, 1972.

Sinkankas, John *Van Nostrand's Standard Catalog of Gems*. New York: Van Nostrand Reinhold, 1968.

Spekke, Arnold, *The Ancient Amber Routes and the Geographical Discovery of the Eastern Baltic*. Stockholm: M. Goppers, 1957.

Srebrodol'skiy, B. I., "Amber in Sulfur Deposits," *Doklady Akademii Nauk S.S.S.R., Earth Science Section* **223**, *1–6:* 204–205. Washington, D.C., 1975.

Stevens, Bob C., *The Collector's Book of Snuff Bottles*. New York: Weatherhill, 1976.

"Diving for Amber," *St. Pauls Magazine,* ca. 1890

Strong, D. E., *Catalogue of the Carved Amber in the Department of Greek and Roman Antiquities*. London: The Trustees of the British Museum, 1966.

Strums, Ed., "Der Östbaltische Bernsteinhandel in der Vorchristlichen Zeit," *Jahrbuch des Baltischen Forschungsinstituts, Commentationes Balticae I*. Bonn, Germany, 1953.

Strunz, H., *Mineralogische Tabellen*. Leipzig: Akademische Verlagsgesellschaft Geest und Portig K., 1966.

Sullivan, Mildred, "The Fashion Outlook," *Modern Jeweler,* Fall 1977, p. 58.

Suprichev, Vladimir, [Amber—Talisman, Medicine, Ornament, in Russian], *Nauki i Tekhnika,* December 1978, pp. 20–22.

Szejnert, Malgorzata, *Traffic on the Amber Route.* Poland, 1977

"Practical Shop Notes," *The Jeweler's Circular,* January 9, 1915.

Treptow, E., *Bergbau und Hütten wesen.* Leipzig: Verlag von Otto Spamer, 1900.

Van Der Sleen, W. G. N., *A Handbook on Beads.* York, Pennsylvania: Liberty Cap Books, 1973

Vargas, Glenn and Martha, *Faceting for Amateurs.* Palm Desert, California, 1969.

Velikiy, N. M., "Amber Finds on the Northwest Shore of the Aral Sea," *Doklady Akademii Nauk S.S.S.R., Earth Science Section* **221**, *1–6:* 164–166. Washington, D.C., 1975.

Villiers, Elizabeth, *The Book of Charms.* New York: Simon and Schuster, 1973.

Walker, T. L., "Chemawinite or Canadian Amber." *University of Toronto Studies, Geological Series,* No. 36, 1934.

Webster, Robert, *Gems, Their Sources Descriptions, and Identification* **I.** London: Butterworths, 1962; and Hamden, Connecticut: Archon Books, 1975.

Webster, Robert, *Practical Gemology.* London: N.A.G. Press, 1976.

Weinstein, Michael, *Precious and Semi-Precious Stones.* New York: Pitman, 1946.

Weinstein, Michael, *The World of Jewel Stones.* New York: Sheridan House, 1958.

Wheeler, Robert Eric Mortimer, *Rome, Beyond the Imperial Frontier.* London: G. Bell & Sons (reprinted by Philosophical Library, 1955).

Williamson, George C., *The Book of Amber.* London: Ernest Benn, 1932.

Yushkin, N. R. and N. Y. Sergeyeva, "Textures of Amber from the Yugorskiy Peninsula," *Mineralogy Doklady Akademii Nauk S.S.S.R., Earth Science Section* **216**; *1–6:* 151–153. Washington, D.C., 1974.

Zaddach, G., *Das Tertiärgebirge Samlands.* Königsberg: Schriften der Physikalishe-Ökonomische Gesellschaft, Jg. 8, 1867.

Zahl, Paul A., "Amber, Golden Window of the Past," *National Geographic* **152**; *3:* 423–435, 1977.

Zalewska, Zofia, *Amber in Poland, A Guide to the Exhibition.* Warsaw: Wydawnictwa Geologiczne, 1974.

Appendix

Care of Amber

Amber beads, rings and other forms of amber jewelry will retain their original luster and splendor indefinitely if a few simple precautions are observed. Because of the softness of amber, its brittleness, and its susceptibility to attack by various chemicals, amber jewelry pieces do require some special care in handling and storing.

CARE OF BEADS

The first precaution is usually taken by the manufacturer by stringing the beads on silk or linen thread with knots between each bead to prevent mutual rubbing and chipping. Knots also provide a safeguard in the unfortunate circumstance of the string breaking, as only one bead will fall, leaving the rest safely knotted on the string.

Amber beads should not be stored where they will rub against metal or other pieces of jewelry. Soft flannel or velvet pouches made with drawstring tops are best for storing and protecting each individual item.

To remove dust and perspiration from amber beads, simply wipe them with a soft flannel cloth dampened with clean lukewarm water. They should then be dried carefully and rubbed lightly with clear olive oil, then rubbed with a soft cloth to remove excess oil and to restore the polish.

Avoid placing amber pieces in any strong solutions, especially ammonia, strong soaps or detergents, and do not allow amber to come in contact with perfume or hairspray. Commercial jewelry cleaning solutions should not be used on amber. All of the foregoing can result in permanently dulling the polish. Furthermore, because some hydrocarbon compounds form a dull white coating on amber, it is advisable to avoid contact with common kitchen substances such as lard, salad oil or butter, and it is a good idea to remove amber jewelry while cooking to avoid excessive heat as well as contact with these substances. Never use hot water for cleaning amber.

CARE OF RINGS

Special care should be taken when wearing amber rings, which should always be removed from the fingers before the hands are immersed in anything but plain water. Avoid bumping rings against any kind of hard surface.

CARE OF CARVINGS AND ART OBJECTS

All such objects should be handled gently to avoid knocks and chipping. Dust may be removed with a soft cloth or feather duster. Avoid placing too near to heating ducts, in direct sunshine or in extremely warm areas. Carvings are best exhibited and protected in glass showcases; however, if showcases are lighted, be sure to have adequate ventilation to avoid excessive heat from the electric light.

REPAIR

Broken articles may be repaired by coating the surfaces to be rejoined with linseed oil, then gently heating the surfaces over a low flame or charcoal fire. When the pieces are warm, press the matching surfaces together and continue to hold the article over the heat until the amber softens.

Epoxy glues may be used for repair of fractured pieces, but great care must be taken to prevent epoxy from coating areas other than the fractured surface. Hardened glue should not be removed from amber with glue solvents, as they will attack the amber.

REMOVAL OF CHEMICALLY-FORMED COATINGS

If amber surfaces have come into contact with alcohol, the resulting white coating can be polished off using water and tripoli aplied with the ball of the hand by rubbing over the piece. If a polishing wheel is used, it is important to use a new felt disc or one that is used only for polishing amber. After use, the polish compound is removed, using cold or lukewarm water.

Golden amber has retained its beauty for 60 million years; with the observance of these simple handling procedures, your amber will retain its beauty indefinitely.

Amber Facts and Fancies*

Amber is a water-insoluble tree resin which has attained a stable state after various changes due to loss of volatile constituents, processes such as oxidation and polymerization, and lengthy burial in the ground. The botanist is not concerned with the length of time to reach this state. It may take a year, a thousand years, a million years or more. The geologist considers amber to be a fossil tree resin, "fossil" meaning evidence of prehistoric life. The resin, then, must have been exuded from the tree at least before recorded history. The purist places even more restrictions upon the term amber. For him "true amber" is that fossil resin containing as much as 3% to 8% succinic acid (hence, known as succinite) and coming from trees which flourished along the shores of the Baltic Sea as long ago as 60,000,000 years.

Gem-dealers and jewelers obviously are purists, since one pays the highest prices for Baltic amber. The geologist works to establish the age of the deposits in which the amber is found. He also attempts to characterize the amber by its physical and chemical properties, and leans heavily upon aid from organic chemists to accomplish this. The botanist, naturally, is interested chiefly in the botanical sources of amber. Interdisciplinary studies are needed to bring the answers to all groups of investigators.

*Reprinted from Frondel, Judith W., "Amber Facts and Fancies." *Economic Botany* **22,** *4:* 371, October–December 1968.

Charts for Comparing Amber with Common Substitutes*

Table of Specific Gravity

1.03 Amber, copal
1.05 Amber, copal, polystyrene
1.06 Amber, copal
1.10 Amber, copal, meerschaum, jet
1.18 Acrylic plastics (plexiglas, lucite, perspex) jet, meerschaum
1.19 Acrylics, jet, meerschaum
1.20 Jet, meerschaum
1.25 Bakelite (amber imitation thermosetting plastic), jet
1.26 Tortoiseshell, jet, bakelite
1.29 Tortoiseshell, jet, bakelite, cellulose acetate plastics
1.30 Tortoiseshell, jet, bakelite, cellulose acetate
1.32 Tortoiseshell, jet, bakelite, cellulose acetate, casein plastics
1.35 Casein plastics, tortoiseshell, jet, cellulose acetate
1.36 Cellulose nitrate plastics (celluloids), cellulose acetate, jet
1.38 Cellulose nitrate, cellulose acetate, vegetable ivory, jet
1.40 Cellulose nitrate, cellulose acetate, vegetable ivory, mineral coal
1.42 Cellulose nitrate, ivory

Refractive Indices

1.49 Polypropylene, acrylic, cellulose acetate
1.50 Celluloid, polyethylene, cellulose acetate
1.53 Nylon, polyethylene
1.539 Amber, glass
1.54 Copal, amber, glass, polyethylene
1.545 Amber, glass
1.55 Casein, bakelite

*Reprinted from *Gemstone and Mineral Data Book* © 1972 by John Sinkankas. Published by Winchester Press.

1.56   Polystyrene-acrylonitrile, casein, bakelite
1.57   Polystyrene-acrylonitrile, bakelite
1.58   Bakelite
1.59   Polystyrene, bakelite
1.60   Ekanite, glass
1.64   Bakelite, glass
1.66   Bakelite, glass

Observations of Recent Resin and Relationship to Amber Origin*

The observation of Recent resin under its natural conditions will provide instructive suggestions concerning the origin of amber. The resemblance of the natural morphological forms of amber with the Recent resin is very striking. In our climate, pine resin is that most strongly resembling the amber.

In full summertide the fresh resin is exuded from the trunks and branches of trees, most abundantly from damaged parts which it stops up and heals. For economic ends the pine trunks are cut deeply, artificially to cause more plentiful exudation so that as much as 10–12 kg of resin may be annually obtained from one tree trunk.

Fresh resin is a gummy, transparent and glittering substance, most commonly of a light-yellow color, resembling fresh July honey. With the coming of the chilly autumn weather the secretion of the resin stops until the next spring season. The current year's incrustations coagulate, solidify and will thus persist, but with time, after a few years, they will grow opaque and without lustre.

In forests where resin is not exploited, we may observe such hardened and tarnished resin secretions, exuded some years ago, side by side with those of the current year, fresh and transparent.

The resin incrustations are variously shaped. The simplest ones are in the shape of small globular drops (the so-called "tears") which, when quite fresh, have the appearance of dew drops. Other resin drops are egg-shaped, sharply pointed at one end, resembling miniature pears or round and flattened like tiny discs.

The size of the resin drops varies considerably. It ranges from that of a poppy-seed to the size of a walnut and even bigger. "Drops" of this size, however, are compound ones, formed in result of a long-lasting, many times repeated exudation of the resin.

The simple (i.e. non-compound) drops are relatively seldom encountered. Stalac-

*Reprinted from Zalewska, Zofia, *Amber in Poland, A guide to the Exhibition.* Wydawnictwa Geologicane, Warsaw, 1974, pp. 2–4.

tites are much more common. Sometimes they loosely hang down the trunk, more often they adhere to the surface of the bark. The simplest forms of stalactites have been produced by one single exudation of the resin. Fairly often we may encounter stalactites made up of several layers of incrustations, secreted in the course of several years, with readily seen boundaries of the successive exudations.

In cases of exceptionally abundant resin secretion, the individual adjacent stalactites may fuse into rivulets or streams flowing along the trunk down to the ground. There, at the foot of the trunk the resin accumulates in lumps varying in size. A such abundant secretion of the resin we can observe in some fragments of the Puszcza Myszyniecka pine forests.

Thus, the natural exudation of the resin most commonly takes on the shape of drops, stalactites, rivulets and lumps. The variability in shape and size is very strong, neither is there a lack of transition forms.

Amber, no matter whether obtained, from the sea, washed out on to the beach by the action of stormy sea waves, or excavated from the ground, occurs in nature under the same forms as the resin of Recent trees.

INDEX